The Lie of the Land

ABOUT THE AUTHOR

John Gibbons was born and raised on a farm in Co. Kilkenny.
The founder of a successful healthcare publishing business,
Gibbons began a second career as an environmental journalist
and activist following the birth of his first child in 2002. He is a
regular contributor to a number of media outlets, including the
Irish Examiner, the *Irish Times* and *The Last Word with Matt Cooper*
on Today FM.

The Lie of the Land

*A Game Plan for Ireland
in the Climate Crisis*

JOHN GIBBONS

SANDYCOVE

an imprint of

PENGUIN BOOKS

SANDYCOVE

UK | USA | Canada | Ireland | Australia
India | New Zealand | South Africa

Sandycove is part of the Penguin Random House group of companies
whose addresses can be found at global.penguinrandomhouse.com.

Penguin Random House UK,
One Embassy Gardens, 8 Viaduct Gardens, London SW11 7BW

penguin.co.uk

Penguin
Random House
UK

First published 2025

001

Set in 13.5/16pt Garamond MT Std
Typeset by Six Red Marbles UK, Thetford, Norfolk
Printed and bound in Great Britain by Clays Ltd, Elcograf S.p.A.

The authorized representative in the EEA is Penguin Random House Ireland,
Morrison Chambers, 32 Nassau Street, Dublin D02 YH68

A CIP catalogue record for this book is available from the British Library

ISBN: 978-1-844-88683-8

Penguin Random House is committed to a sustainable future
for our business, our readers and our planet. This book is made from
Forest Stewardship Council® certified paper.

MIX
Paper | Supporting
responsible forestry
FSC
www.fsc.org
FSC® C018179

To Jane, Sophie and Simone, with love and gratitude.

Contents

1. Goldilocks Is Dying

Earth has long been described as a Goldilocks planet, with climatic and atmospheric conditions uniquely suited to life. If Earth is such a planet – not too hot, not too cold – then Ireland, with its temperate climate, is a Goldilocks country. By an accident of geography, Ireland exists in a climatic sweet spot, sheltered from the very worst extremes of heatwaves, hurricanes, droughts and deluges.

But Goldilocks is dying. The long era of uncanny global climatic stability has ended; the climate has changed more radically in the past half-century than at any other time since the end of the last Ice Age. Levels of atmospheric carbon dioxide (CO_2), the planet's key heat-trapping gas, are increasing around a hundred times faster than would be the case under natural warming conditions.[1] Once a molecule of CO_2 is released, it continues to warm the atmosphere for between 300 and 1,000 years.[2] For every litre of fuel burned, the long-term warming effect is 100,000 times greater than the actual heat released at the moment of combustion.[3] That's the insidious power – and danger – of greenhouse gases. When rapid increases in other greenhouse gases such as methane and nitrous oxide are also tallied, the picture becomes even darker.[4] As a result of the proliferation of greenhouse gases, the planet is warming ten times faster than at any time in the last 65 million years.

In 2024, global average temperatures were 1.55 degrees Celsius above the pre-industrial average.[5] That made it the hottest year on the instrumental record and, in all likelihood, the hottest in over 120,000 years. The second hottest year ever recorded was 2023. And the eleven years 2014–2024 were all warmer than any year on record pre-2014.

The water that laps along Ireland's shorelines is warmer and more acidic now than it has been in millennia. And, although you would not be able to detect it when you inhale, the very air that we breathe has also changed. Today, it contains some 50 per cent more CO_2, and 270 per cent more methane, than in the pre-industrial era.[6] Incredibly, a single species has, in just a few generations, altered the very chemistry of the Earth's atmosphere.

The distinct interglacial climatic era known as the Holocene, which prevailed for roughly the last 10,000 years, was a period of quite remarkable climatic stability. It was thanks to the conditions of the Holocene that our ancestors were able to forgo eons of nomadic hunter-gatherer existence, settle down and use the land to produce the food surpluses that gave rise to human civilization.

The Holocene, although still the current geological epoch, has now effectively been supplanted by an era scientists identify as the Anthropocene, so named because humans are now the most impactful drivers of planetary change.[7]

In physical size and human population, Ireland may appear globally insignificant. The future of human civilization will not be determined here. But the story of Ireland's relationship to the climate crisis – the story that this book will tell – is a fascinating microcosm of global dynamics. And the people of Ireland face a choice – or, more accurately, a spectrum of choices that will largely determine what life here will look like in the age of global heating.

The bad news – there is no avoiding it – is that grave

damage has already been done, both globally and in Ireland, and that much of it is irreversible. The good news is that there is still much to play for. Climate action is often described in terms of humans making sacrifices in the short term to secure a long-term future, but this framing is now out of date. With each passing year it becomes ever clearer that, barring a swift change of direction, most humans currently alive will see the world change dramatically for the worse.

*

Like almost every other country in the world, Ireland has seen a sharp rise in extreme weather events in recent years. There is a general understanding that these events are linked to climate change, and are only going to get worse. But for the vast majority of people in Ireland, the climate crisis remains a peripheral issue, like the sound of distant thunder, breaking to the surface of consciousness every now and again, then quickly ebbing away. The great majority of people are, it seems, more concerned about the end of the month than the end of the world.

I can say this with some confidence and empathy because, for almost the first forty years of my life, I was one of those people. I recall in my youth reading about rainforests being cut down and whales being hunted, but to my parochial outlook, these were things happening in remote parts of the world that had absolutely nothing to do with me.

I grew up on a farm in rural Kilkenny, the tenth of twelve children. My late father, Michael, was a driven man, and was regarded by his peers as a progressive and successful farmer. He bought a large arable farm in the 1950s and set about draining and improving the land over the following two decades.

One field, more than forty acres in size, was known as the

'big seven': it had been created by turning seven smaller fields into one. Though I was oblivious to it at the time, hedgerows are one of the few remaining refuges for wildlife in rural Ireland. Turning seven fields into one involved destroying the hedgerows that had separated them, and this involved the loss of habitat and biodiversity.

I made a personal contribution to this on one occasion when a JCB had been hired to dig out a mature hedgerow. As it worked its way up the field, rabbits and other small animals staggered out of the wreckage, some injured, all disoriented. They were swiftly dealt with by bystanders armed with hurleys, this author included.

The farming life that I grew up in was a tough, low-margin and unsentimental business. What I learned from my father is that there are two types of animal: livestock and vermin. Badgers, rabbits, crows, foxes – all were vermin and thus fair game for extermination. I was regularly dispatched, shotgun or rifle in hand, to blast anything that moved. In spring, we targeted the nests of crows. Pigeons, foxes and rabbits were shot all year round, and badgers killed when encountered, which thankfully was extremely rare. Other people will, of course, have had a very different experience growing up in rural Ireland in those years and subsequently, but at no time did I have any sense that our approach to wildlife was particularly unusual or atypical. There were moments, like when the early morning dew draped itself across a hundred spider webs, that you had a momentary sense of the unfolding miracle and beauty of life all around you, but these were very much the exception.

In the local Christian Brothers primary school, I was one of only a handful of culchies in a classroom of around fifty boys, and regularly found myself cast in the role of chief defender of farmers and rural Ireland to a largely hostile audience that sometimes included the teachers. The prevailing narrative was that us boys from the countryside were backward, unsophisticated and a bit thick. There is nothing really new about the culture wars that surround Irish agriculture these days. Then, as now, people were often more inclined to imagine themselves in mutually hostile camps than to wonder who might be stirring up this animus in the first place.

*

My first child, a daughter, was born in late 2002. This was a watershed moment, in ways I could scarcely have anticipated. Almost overnight, my temporal perspective shifted. The furthest future horizons were no longer measured in weeks and months but in years and decades. I found myself drawn to explore what Ireland and the wider world might look like when my daughter reached middle age. Nothing in my life up to that point could have prepared me for what I was to discover.

This journey began serendipitously, when I came across a copy of *Something New Under the Sun: An Environmental History of the Twentieth-Century World* by John R. McNeill of Georgetown University. The human race, he wrote, 'without intending anything of the sort, has undertaken a gigantic uncontrolled experiment on the Earth'.

McNeill calculated that, in a single century, humans employed more energy than in all of our history up to 1900. This massive spasm of activity saw world population quadruple, the global economy grow fourteen-fold, energy use increase thirteen-fold, water use expand nine-fold, cattle population grow four-fold and marine-fish catch increase thirty-five-fold. Most impactfully of all, CO_2 emissions increased seventeen-fold. My maternal grandmother was born in Kilkenny in 1901. Over the brief three-generation span from her life to mine, humanity has reshaped an entire planet.

My father died suddenly one Saturday in late 1989. While our relationship had always been tense and sometimes fractious, the event threw me into a state of shock that persisted for weeks. I had almost forgotten that feeling until it returned with a vengeance nearly fifteen years later, after I had read, then reread, McNeill's book. For weeks, even months afterwards, I found myself in a state that cycled through shock, disbelief and denial. Initially, I found it impossible to accept that the situation could possibly be as bleak as it appeared. After all, at that time there was little or no sustained media coverage of the climate and ecological crisis. If this was indeed real, how could it be possible that nobody was talking about it? But the more books I read, the more I realized just how much trouble we were in.

Every five or six years since 1990, the United Nations Intergovernmental Panel on Climate Change (IPCC) has published a comprehensive Assessment Report. The third of these was released in 2001. It stated that, since the mid-twentieth century, most of the observed warming of the planet was 'likely' due to human activities. The report added that climate change would have 'beneficial and adverse effects on both environmental and socioeconomic systems, but the larger the changes and the rate of change in climate, the more the adverse effects predominate'.

With such wishy-washy language, it is no wonder the 2001 IPCC report made little impression. It certainly passed me by completely. By the time the next report was released, in 2007, it was a different world. And this time, I was actually paying attention. The earlier ambiguous language was supplanted by a sense of real urgency. Among the report's findings was that 'warming of the climate system is unequivocal'.[8] That might seem painfully obvious today, but 2007 was the first time the intensely conservative IPCC actually spelled it out. It was also the first time the IPCC acknowledged the very real possibility of climate disaster ahead, stating that 'unmitigated climate change would, in the long term, be likely to exceed the capacity of natural, managed and human systems to adapt'. You didn't have to read too closely between the lines to realize this careful language was signalling the very real possibility of widespread system collapse.

A year earlier, the release of *An Inconvenient Truth*, a documentary film by former US vice-president Al Gore, had tapped into growing public unease about shifts in the climate system. The catastrophic flooding that inundated New Orleans in the wake of Hurricane Katrina in 2005, with nearly 1,400 fatalities and damage costing over $100 billion, provided a visceral backdrop to the film. By the time I saw it, I was already well up to speed with the science, but I was still a bystander. The intensely moving experience of watching *An Inconvenient Truth* provided at least some of the emotional impetus I needed to start taking some action.

I first met John Sweeney of the ICARUS climate unit in Maynooth University that year. An affable and approachable Glaswegian who has lived and worked in Ireland since the late 1980s, Sweeney teased out the science and patiently answered my many questions. That one-on-one encounter left me in no doubt as to the gravity of the climate crisis and redoubled my desire to get involved.

During 2007 I launched a blog, ThinkOrSwim.ie, exploring climate, environment, energy and related issues. I had heard on the grapevine that the *Irish Times* might be on the lookout for climate-related contributions, so took a chance and pitched the opinion editor, Peter Murtagh. On the back of this, he asked me to contribute an opinion article. It was published in mid-March 2008, and to my complete surprise this grew into a weekly column, covering the spectrum of climate and ecological issues.

I got a first real inkling of what I was letting myself in for in early 2008, when I visited California to meet Stephen Schneider, a renowned climatologist, at his office in Stanford University. I was required to supply multiple forms of identity before being granted an interview. In person, Schneider was apologetic about all the security, explaining that the FBI had informed him that his name was on a neo-Nazi 'death list'. He received hundreds of abusive emails every week and his home phone number and address had been delisted. He had also installed an elaborate home security system. His crime? Speaking out publicly about climate change.

I now realize I was quite naïve about the nature of the 'climate debate' in the media and beyond at that time. My encounter with Dr Schneider should have alerted me to the fact that it was less about science communication and more a battleground of ideology and powerful vested interests. I innocently assumed that once people were presented with the scientific facts in a clear, consistent and logical way, then public opinion would shift decisively in favour of strong climate action. As I was to discover, this is not how the world works.

With two young children at home and a demanding full-time job running a business that was entirely unrelated to this time-sapping and emotionally draining new journalistic commitment, I began to wonder if I hadn't made a terrible mistake. And then, in the bitterly cold January of 2010, I received an email from the *Irish Times* with the subject line 'Bad news'. My weekly column was getting the chop.

The COP15 international climate conference in Copenhagen the previous month had just ended in failure and disarray. It was derailed at least in part by a clever hoax known as 'Climategate', which involved the cherry-picking of two or three phrases from a cache of tens of thousands of hacked emails between climate scientists to create the entirely false impression that scientists were conspiring to tamper with the data to exaggerate global warming.[9] This was shown to be groundless, but not before headlines about the supposed 'scandal' went around the world. The hoax was crude yet effective, as it depended for its success on the limitations of the public's – and the media's – understanding of how the scientific process actually works.

Meanwhile, as I was still coming to terms with the axing of my column, an editorial appeared in the *Irish Times*. It was mostly about practical issues arising from an extended spell of cold weather in Ireland, but its headline ('Global Cooling') was shockingly misjudged, as were its opening sentences: 'So much for all of that guff about global warming! Are world leaders having the wrong debate? We are experiencing the most prolonged period of icy weather in forty years and feeling every bit of it.'[10]

I have no way of knowing whether or not the cancellation of my weekly column was connected to Climategate or the cold snap. But it was probably indicative of the thinness of the soil in which editorial understanding of the climate crisis was sown. Either way, I was more relieved than disappointed.

*

There should probably be a health warning for anyone who tries to wrap their head around the climate and biodiversity

9

emergency. It bruises you deeply, in ways that are difficult to describe adequately. I have been peering into this abyss as a journalist, as a parent and as a human being for more than twenty years, and it still leaves me numb and shaken. Nothing in my previous life ever came remotely close to preparing me for this. Mark Twain once described education as the path from cocky ignorance to miserable uncertainty, and that about sums up my own experience over the last two decades in coming to terms with the reality of the global climate emergency.

As a species, our system of reasoning by mental shortcuts is optimized for making quick decisions based on the best available information or hunch. This system is ill equipped to respond to a slow-moving or seemingly distant hazard. We also are prone by nature to heavily discount future costs against even modest present gains. The philosopher Timothy Morton coined the phrase 'hyperobject' to describe something of such vast temporal and spatial dimensions that it defeats traditional means of thinking about it.[11] The ecological emergency is the ultimate hyperobject, overwhelming our senses and rendering itself both omnipresent and near-invisible, as if the very light by which we might perceive it were emanating from beyond the visible spectrum.

Many in the climate science community are now acknowledging that the implications of the work they are doing are leaving them with what could be called 'pre-traumatic stress disorder'. They can clearly see the ecological and climate tsunami sweeping towards us, but humanity is, figuratively speaking, relaxing on the beach as the wave draws ever closer. The world we once knew, the world our parents were born into, simply no longer exists, and it is never coming back. Coming to terms with that stark reality is no small

undertaking. The understanding that the world is growing more and more hostile to life is not easy to truly take
on board or to communicate to others. To save lives and
mitigate ecological damage as the climate emergency deep-
ens, we must first abandon false hope, the sort of hope that
is itself the enemy of resolve.

What is needed, psychologist Eoin Galavan told me, is 'a
balance between our capacity to bravely, creatively and con-
structively meet challenges around the climate crisis and the
recognition of the peril that we are in and the damage we've
already done'. The Covid pandemic provided a case study in
what happens when there is such an overwhelming focus on
a clear and present danger that many of the usual political
and social assumptions are reconsidered. 'I don't remember
people saying should we be optimistic or pessimistic about
how we talk about Covid; we just said it's dangerous, it's
going to kill us, so here's what we have to do to protect ourselves,' Galavan said.

American journalist David Wallace-Wells, who had no prior
background in climate reporting, caused an international
sensation in 2017 with the publication of an extended essay
titled 'The Uninhabitable Earth'.[12] Researching the article,
and the book that grew out of it, left Wallace-Wells finding it
'impossible to even consider our likely future without recoil-
ing in horror and grief'.

When I interviewed Wallace-Wells in 2020 he explained
the secret to his relatively upbeat approach to his prediction.
'One of the reasons I've been able to work on this material
is that, as a journalist, I keep a kind of a distance from it. I
often joke it's as if all journalists are somewhat sociopathic
in their ability to regard stories as just stories.'

For Wallace-Wells, being able to use his platform to warn

his readers is also a personal coping mechanism. 'It's the unique privilege of being a journalist; when I'm feeling especially distressed about the future of the planet, I tend to think about what I can do.'

Psychologist John Sharry told me: 'The silver lining of eco grief is that for many people it makes them more sensitive, more tuned in and more appreciative of the life they have right now, and to savour their relationships with their children and friends.' Realizing how fragile and precious our lives really are 'leads many people to live their lives more meaningfully', he added.

After giving talks or presentations, I'm often asked why I don't strike a more optimistic tone. What about renewable energy, solar panels, heat pumps, cattle feed additives and electric vehicles (EVs)? Surely they are tangible evidence that we're finally getting on top of this crisis?

The numbers, alas, tell an altogether different story. The intergovernmental process to tackle climate change has been underway in earnest since around 1990. In the three and a half decades since then, we have released more heat-trapping greenhouse gases (GHGs) into the global atmosphere than in all of human history prior to 1990.

Renewable energy is indeed growing fast, but so far it is simply being used to serve our ever-expanding and seemingly insatiable energy demands, rather than to reduce fossil-fuel usage. In 2024, amidst the ongoing revolution in wind and solar energy, humans burned more fossil fuels than in any previous year.[13]

*

Ireland's Climate Action Plan commits us to achieving a 51 per cent cut in total GHG emissions by 2030, while the separate but related EU Effort Sharing Regulation gives

Ireland a binding emissions-reduction target of 42 per cent by that year. Based on the lack of progress to date, these objectives remain completely out of reach. In mid-2025, the Environmental Protection Agency (EPA) reckoned the absolute best-case scenario for emissions reductions is now 23 per cent. Even this seems fanciful, as it would require an improbable 'full implementation of a wide range of policies and plans across all sectors and for these to deliver the anticipated carbon savings'.[14] And in March 2025 a report by the Irish Fiscal Advisory Council and the Climate Change Advisory Council (CCAC) warned that Ireland is on track to fall far short of EU-mandated targets in four separate areas by 2030, and will be on the hook for between €8 billion and €26 billion in fines if it continues with business as usual. Ireland currently ranks last of the twenty-seven EU states in terms of meeting its Effort Sharing obligations.[15]

Clearly, we are on the wrong path. In this book, we will look at where Ireland stands today, and we will explore how the necessary changes can be brought about.

It is no accident that I devote three chapters to agriculture and food security. Agrifood are the source of almost 38 per cent of total national emissions – by far our largest single contributor, and the sector that has been the most resistant to doing its fair share on emissions reductions.[16] One might imagine that the GHG footprint of Irish agriculture is roughly proportionate to the sector's importance in the national economy. The truth is very different. Ireland's agri-food industry, which includes farming, fisheries and food processing, employs 6.4 per cent of the national workforce, and accounts for around 6.7 per cent of gross national income and 9 per cent of exports.[17] These are significant numbers, but nowhere nearly proportionate to its negative environmental impacts.

Because of our oversized agricultural emissions, mainly emanating from our huge livestock herd, Ireland has the third highest overall emissions per capita in the EU. A 2017 study found Irish agriculture to be the least climate-efficient in the entire EU.[18] This sits most uncomfortably with the dominant narrative that Ireland's largely grass-based live-stock systems are inherently or uniquely 'greener' than those of other countries.

We need a new narrative and new ways of operating for agriculture – and also for transport, housing, and energy generation. Our task is two-fold: to transition away from fossil fuels as quickly as possible while also investing heavily in building resilience against the climatic and socio-political upheavals that are already baked in. The range of possible outcomes for Ireland is wide: there is everything to play for. For all its challenges, Ireland, with its Goldilocks micro-climate, is one of the very best locations in the world in which to weather the coming climatic storm.[19] To capitalize on our natural advantages while shoring up our many vul-nerabilities will require this generation to make some tough political choices. Should we succeed, the prize of a resilient, energy-independent and food-secure future is within our grasp. Failure cannot be an option.

2. The Lie of the Land

The farm that I grew up on in the 1960s and early 1970s was typical of the time. The farming system was mixed – a combination of beef cattle, sheep, cereals and vegetables, managed in rotation to give the land a chance to recover and to recycle animal manure as efficiently as possible.

The use of costly inputs, such as fertilizers, feeds or pesticides, was very limited. We kept a cow to provide milk for the house, and a pair of pigs to eat household food waste. There were hen houses in the haggard, from which we collected fresh eggs every morning and, every now and then, a chicken for the pot.

In the kitchen garden, fruit and vegetables were grown for our own consumption. As a small child, I regularly trailed out into the kitchen garden in the early evening as my parents and an older sibling or two tended to the patch after the day's farm work was completed.

In hindsight, it was a bucolic picture, and it is probably pretty close to many people's idea of what life is like on a typical Irish farm today. But the truth is that, even during my childhood, that way of life was already dying out. Our kitchen garden was abandoned in 1970. The arrival of relatively cheap vegetables in the local shops meant my parents felt it was no longer worth the effort of growing our own. We moved to a larger farm on the other side of Kilkenny city two years later, and that was the end of the household cow and chickens, though we still kept a pair of pigs.

The change in our family farming circumstances coin-cided with wider shifts in agriculture. In 1973, Ireland joined the European Economic Community (EEC), the forerun-ner of the EU. Irish farmers were now operating within the EEC's Common Agricultural Policy (CAP), which meant guaranteed prices and rising farm incomes. The new farming era would be defined by specializing in a single product and ramping up production.

Half a century later, our rural landscape has been transformed. In 1973, one in four Irish people worked in agriculture.[1] Today, it is fewer than one in twenty-five. While the numbers working in agriculture have plunged, total production has dramatically increased, bringing with it huge benefits for some, but at a fearsome cost in terms of GHG emissions, biodiversity loss and plummeting water quality.

It didn't have to be like this. The ramping up of agricul-tural activity after we joined the EEC initially led to increased emissions and marked declines in water quality. However, by the early 2000s, GHG emissions and water pollution from agriculture were declining steadily, as a direct result of EU-directed policies.[2]

There were two reasons for this progress, according to a Teagasc study: 'declining ruminant livestock populations and declining inputs of manufactured fertilizers'. This followed changes in the CAP aimed at reducing beef cattle and sheep numbers, and was also assisted by EU limits on dairy pro-duction. Further reforms in the early 1990s saw major cuts in price support for arable crops and beef, offset by com-pensation payments, a scheme that was extended in 2003 into the single farm payment, decoupled from production.[3] This regime saw emissions from Irish agriculture fall from a

1998 peak of around 22 million tonnes of CO_2-equivalent to around 18 million tonnes by 2011.

Had this pathway been maintained, then the whole sector would have been primed to meet the more ambitious climate targets coming down the line at both national and EU level. But in 2008 the EU decided to lift its milk quotas – essentially limits on national production. The limits would cease to apply from 2015.[4]

The Irish general election of early 2011, meanwhile, was to mark a decisive shift in direction for agriculture. Emerging from the ashes of the economic crash of 2008, the new Fine Gael-led government was casting around for home-grown options to stimulate the moribund Irish economy, and with milk quotas due to be lifted, dairy seemed to fit the bill. A template for an expansion of the dairy sector already existed, in the form of an industry-developed strategy document titled 'Food Harvest 2020', published by the Department of Agriculture in 2010.[5] The following year, shortly after becoming Minister for Agriculture, Fine Gael's Simon Coveney wrote the Foreword to the first annual progress report on the strategy. If there was any doubt that government and industry were in lockstep, Coveney laid it to rest. 'I am very aware that many of the steps needed to be achieved by 2013 and 2015 are primarily commercial decisions for industry,' he wrote, 'but I hope that the industry will work with me and the State Agencies to meet, and indeed exceed, the ambitious targets.'

In what would become a familiar pattern, 'Food Harvest 2020' was heavily cloaked in the language of faux sustainability. It noted the historic link between Ireland and the colour green, adding that associating itself with this colour, and its connection in the public mind with concern for the environment, would lead the industry's overseas customers to recognize that, by buying Irish, 'they are choosing to value and respect the natural environment'.

The rewards of dairy expansion were tantalizing, but the risks were obvious. An EPA analysis of 'Food Harvest 2020' concluded that the strategy it outlined would lead to a 7 per cent increase in overall agricultural emissions driven by 'a projected increase in the national herd'.[6] Teagasc economist Trevor Donnellan wrote in a 2009 paper that to meet Ireland's EU-mandated 20 per cent GHG reduction target by 2020, 'even with reduced fertilizer usage and more extensive production practices, a very substantial decrease in the livestock population is required'.[7] In the same paper, Donnellan and colleagues concluded that 'agricultural policy is likely to contribute to a reduction in GHG emissions from agriculture over the next decade'.

The exact opposite happened. This detailed Teagasc analysis, which pointed to the necessity to sharply reduce beef cattle numbers in order to offset emissions from the expanding dairy sector, failed to find its way into the government's published plans. Teagasc seems to have chosen not to back its own analysis, arguing instead that various technological, dietary and other tweaks could somehow achieve the same effect. In any case, all the key stakeholders understood the reality: increased stock levels would drive higher emissions. As the head of the Irish Cattle and Sheep Farmers' Association, Gabriel Gilmartin, noted in 2012, 'On the one hand, "Food Harvest 2020" says we must strive to dramatically increase agricultural production – but on the other, we are expected to reduce the emissions from farming activities. The two targets are plainly incompatible.'[8]

Prior to the 2011 election, Simon Coveney had been Fine Gael's environment spokesman. I had in-depth discussions with him in this capacity, and was impressed by his grasp of the environmental brief. I attended an event in Dublin in January 2008 at which he excoriated the then Fianna Fáil–Green government for failing to set sufficiently stringent emissions targets, and spoke movingly of having had his 'own worst fears confirmed' as to the scale and severity of the climate crisis.

However, on Coveney's appointment as agriculture minister in March 2011, all those fears seemed to evaporate as he embraced industry expansion plans with gusto while brushing aside environmental concerns. In September 2013, he disparaged the European Commission's climate targets for the agriculture sector, claiming they made 'no sense to me on any level'.[9] He uncritically adopted the industry's dubious assertion that the Irish milk and beef sectors had among the lowest carbon footprints in Europe, so that limiting them would simply benefit a more emissions-intensive form of production taking place elsewhere. (We will interrogate both those claims in more detail later.) In a newspaper interview, Coveney stated that he had clashed with the EU Commissioner for Climate Action, Connie Hedegaard, telling her forcefully that the EU 'has got it wrong in relation to the food industry'. He added that, with Ireland producing an astonishing 10 per cent of the world's infant formula, the government had no intention whatever of limiting this fast-growing sector to comply with climate targets.

In November of the following year, Coveney stated the government's philosophy in the bluntest terms possible at the National Dairy Conference in Dublin: 'I will not allow a situation where the potential for growth and expansion in agri-food will be compromised by the setting of emissions limits.'[10] Further, Coveney told a meeting of the Irish

Farmers' Association (IFA) he would be an 'aggressive and proactive advocate' for continued agricultural intensification. This was a remarkably brazen approach for a minister to adopt – akin to a transport minister telling motorists they should not have their journey time extended by having to comply with the legal speed limits.

The febrile atmosphere ahead of the lifting of EU milk quotas in April 2015 was captured by Rabobank Europe's analyst Kevin Bellamy, who said the mood in the dairy sector was 'like the night before Christmas'. The banks poured hundreds of millions into investment loans to fuel the ramping up of dairy operations.[11]

The following year, during an RTÉ *Prime Time* debate, Coveney claimed that the national dairy herd would expand by an additional 300,000 cows over the coming five years 'while maintaining the existing carbon footprint of the agriculture sector'. There was no robust evidence available, then or now, to support his contention that such a dramatic increase in dairy cow numbers could occur without a commensurate increase in emissions of all kinds.

If the Minister for Finance were to appear at a conference of bankers and promise to be an aggressive lobbyist for their growth ambitions, even if these ambitions were contrary to the government's own definition of the public good, there would probably be a political outcry. However, it has become completely normalized for the Minister for Agriculture and the department he or she heads to be uncritical advocates of the industry's plans.

Coveney was replaced as agriculture minister by his Fine Gael colleague Michael Creed in May 2016. The personnel change had no impact whatever on policy. A new storyline was being developed by the sector to magic away emissions, and this involved the concept of 'decoupling': the idea that the link between production increases and rising emissions had been decisively broken.

In April 2018, Creed told the Dáil that, between 2012 and 2016, dairy cow numbers had risen by 22 per cent and milk production by 27 per cent while emissions in the same period rose by 'only' 8 per cent. This, the minister stated, 'demonstrates [that] a level of decoupling is occurring'.[12]

Creed's statement was completely misleading, because the emissions figure he cited related to the entire agriculture sector – as he later had to admit when pressed on the matter. There was no 'decoupling': dairy-related emissions increased by 25 per cent in the period, in lockstep with the growth of the dairy sector.

I wrote about the government's sleight of hand for the investigative website DeSmog,[13] and the London *Independent* newspaper ran a news item titled 'Irish Government Using Wrong Data to Downplay Greenhouse Gas Emissions from Cows' that drew on my report.[14] But the story attracted little or no media attention in Ireland.

It would be more than five years later, in December 2023, before anything closer to the full picture would emerge. This came in a detailed statement by the Department of Agriculture's chief inspector, Bill Callanan, to the Oireachtas Joint Committee on Climate Action. In his concluding remarks, Callanan said the department 'acknowledges the accumulated negative impacts on biodiversity and climate which our sector has contributed to through lack of or misinformed action, alongside other sectors over past decades'.[15]

Callanan's remarks dovetailed with the devastating observation made just a couple of months earlier by former EPA director Micheál Ó Cinnéide that 'vested interests in a narrow room' were controlling Ireland's agri-industrial policy. The environmental harms that would flow from the policies set out in 'Food Harvest 2020', Ó Cinnéide told an Oireachtas committee hearing, should not be 'a surprise or a shock'.[16]

*

Though it seems hard to believe today, farmers and their representative bodies were once viewed by the authorities as 'pipsqueaks' (Charles Haughey's memorable phrase) and troublemakers.

As recently as the early 1960s, almost 40 per cent of the Irish population was still engaged in agriculture, which comprised by some distance the largest part of the national economy. In 1962, the National Farmers' Association (NFA) demanded a package of subsidies totalling £83 million, stating this was necessary to bring the income of farmers in line with wider society. (This was, of course, before the EEC began subsidizing Irish farmers through the CAP.) At the time, the Department of Agriculture rejected this claim outright, arguing that 'the NFA analysis of the situation is very faulty indeed, and their requests are most extravagant'.[17]

This view was echoed by the Department of External Affairs, which argued further that such efforts at artificially levelling income would involve 'transforming our economy into something approaching a totalitarian system, where the State is all-powerful and the individual secondary'.

The farming community had an especially fraught relationship with Seán Lemass, who served as Taoiseach from 1959 to 1966. Lemass's government was intensely annoyed that, in setting out its demands, the NFA failed to mention that the Irish state was already spending heavily on agricultural subsidies. These had risen sharply, from £26 million in 1960–61 to £39 million in 1962–3, accounting for over 15 per cent of total exchequer spending.

Lemass explicitly warned the NFA against 'political strikes' designed to pressure the government. The response to such actions, he said, would be 'strong measures'. This is all a long way from today's ultra-cosy relationship between farm leaders and the state, and the vice-like grip of agri lobbyists on the Department of Agriculture.

Relations between the NFA and the government hit rock bottom on 24 April 1967. That evening, in a special broadcast on RTÉ television, Taoiseach Jack Lynch threatened to have the NFA proscribed as a criminal organization, essentially placing its membership on a similar footing to the IRA. This extraordinary political intervention came after some six months of protests and campaigning against the payment of agricultural rates.[18]

'By their speeches and actions,' Lynch said, 'the NFA leaders have shown they are prepared to challenge the basic political institutions of this country.' That morning, police had raided a number of farms belonging to NFA leaders.

The front page of the following morning's *Irish Times* is now something of a family heirloom, as it ran a photo featuring my parents and five of their children, including me, the tearful youngest, outside our farmhouse. The same photo also appeared that morning in the *Irish Independent*, captioned 'A Sad Moment'.

Our farm had been the first to be raided at dawn that Monday morning, with a massive show of force. My father, Michael Gibbons, was a leading figure in the NFA and had been involved in the organization's extended campaign of protests, including commodity strikes and a farmers' march on Dublin. He was one of nine NFA men who had engaged in a gruelling three-week sit-in on the steps of the Department of Agriculture the previous winter.[19] Newspaper reports put the number of gardaí at our farm that morning at around 150, backed up by thirty Special Branch officers. My father was first to wake, and from his bedroom window saw the front lawn swarming with uniformed men and at least two 'Black Maria' vans.

There was banging on the front door and my father opened it to the county sheriff, backed up by gardaí. The sheriff demanded payment of rates arrears amounting to £915. When my father refused, gardaí entered the house and began to seize anything that wasn't nailed down. Among the items taken were our car, a washing machine, electric heaters, a radio and a lawnmower.

Meanwhile, more officers were in the farmyard making repeated attempts to drive our small tractor up the ramp of a lorry. I had a bird's eye view of the drama that morning, and although I was not yet four, the memories remain vivid. When our phone in the hallway rang, my father attempted to answer it but was physically obstructed by a garda, despite my loud intervention to the effect of 'that's Daddy's phone'.

By a curious quirk of fate, my father's only brother, Jim, was on the other side of this conflict: he was a junior minister in the Fianna Fáil government at that time. He would go on to became defence minister and end up testifying against Charles Haughey in the bitter Arms Trial in May 1970. Later, as agriculture minister, in 1973, Jim Gibbons was lead agriculture negotiator on Ireland's entry into the EEC. While farming within a few miles of one another in Co. Kilkenny, the two brothers were not close, and the events of April 1967 likely deepened the divide between them.

Two other Kilkenny farms owned by prominent NFA

members were also raided by the gardaí that morning, while the day saw a total of nineteen farmers jailed for their part in an earlier NFA road blockade.

Ultimately, in the showdown between the government and the NFA, the government blinked first. By 1969, the government had formally accepted the NFA's right to negotiate on behalf of farmers. At the same time, the NFA – which in 1971 merged with a number of smaller organizations and was renamed the Irish Farmers' Association – was handed a role in shaping the Programme for Economic Expansion. The association went on to be the undisputed fulcrum of power in rural Ireland and a highly influential player in shaping national policy.

In the first six years of Ireland's EEC membership, agricultural exports trebled, commodity prices rose by 160 per cent due to EEC price support measures, and factory prices for beef cattle doubled.[20] Overall, farm family incomes, many of which had been below the poverty line in the 1960s, more than doubled, and by 1978 average farm income was 15 per cent higher than pay in manufacturing, having been barely half of it just twelve years earlier. It was owing to this combination of events that, by the end of the 1970s, it was said that the three most powerful organizations in Ireland were the Catholic Church, the GAA and the IFA.

Over time, unchecked power can lead to hubris, as was seen in the Catholic Church, and so it was also to be for the IFA. I remember being quite shocked to read the document published by the association in 2015 to celebrate its sixtieth anniversary. It was titled 'The Path to Power', as though the IFA saw itself on a par with a major political organization.[21]

In a year that should have been a triumphant celebration of six decades of success, 2015 was instead to become the

IFA's *annus horribilis* as it stumbled from controversy into full-blown crisis. That year, the association was rocked to its foundations with the revelations that its general secretary, Pat Smith, had received pay and benefits worth some €500,000 a year in 2013 and in 2014.

For good measure, Smith's severance package on exiting his post was valued at €2 million. To put these numbers into context, the typical farm family income at that time was around €24,000. Every bit as hard for ordinary farmers to stomach was the further revelation that IFA president Eddie Downey was receiving close to €200,000 annually, or some eight times the average farmer's income, for what is a demanding but essentially part-time job.[22]

*

The IFA's transformation into an organization that Irish governments respect and fear helps explain the context in which the environmentally ruinous plan to rapidly intensify the dairy sector was hatched. Where once they regulated the sector, politicians are now largely consigned to the role of cheerleading and funding industry plans. Another strand of that story is the spectacular rise of the billion-euro agribusinesses as powerful players in shaping agri-industrial policy in Ireland.

It is ironic that most of the big Irish agribusiness corporations – hugely profitable capitalist operations – have their roots in the co-operative movement, which had a profoundly different ethos.

In 1894, Anglo-Irish reformer Horace Plunkett founded the Irish Agricultural Organisation Society (IAOS). (In another twist of history, my father was president of the Irish Co-operative Organisation Society (ICOS), as the IAOS was renamed in 1979, at the time of his sudden death in

1989.) Plunkett saw that farmers were getting a raw deal: they were paying too much for inputs while not getting enough for their produce; and they suffered from a lack of access to credit.[23] The co-operative movement offered, in Plunkett's words, 'a means to create social cohesion and provided a platform for small farmers to contribute more effectively to national development'.[24]

The Irish rural tradition of *meitheal*, whereby neighbours and extended family groups pitched in to help one another at busy times, had been in sharp decline since the Famine. The co-op movement sought to rekindle this latent co-operative spirit.

It was also very business-minded. As early as the 1880s, Irish agricultural exports were coming under pressure from foreign competition. The dominance of Irish butter in the British market at that time was undercut by imports from Denmark, where co-operative creameries gave the Danes a decisive competitive edge that reformers like Plunkett sought to emulate. For much of the twentieth century, dairy co-ops operated in almost every parish in Ireland.

Direct government supports for dairy expansion were first introduced in the 1960s. A 1963 government report recommended the amalgamation of existing dairy co-ops into larger groups in order to be able to achieve the economies of scale and specialization. By 1997, two big dairy groups, Avonmore (which had subsumed our tiny local creamery in Kilkenny) and Waterford, merged to form Glanbia PLC, which was at the time one of the world's largest dairy processors.

The original ethos of the co-ops was to protect the smaller producer. As the dairy sector evolved, though, this ethos gave way to a different set of values. In an analysis of the sector, Proinnsias Breathnach of Maynooth University writes: 'All the evidence is that most Irish dairy co-operatives have either encouraged, or acquiesced in, the elimination of the small producer from the industry.' The remuneration of the new managerial class that ran the co-ops was tied to performance, and the advent of private shareholders established profit as the primary objective of these organizations. Further, the momentum to privatize these co-operatives 'emanated from the dairy groups' professional management', which meant that 'share price and profitability replaced the milk price as the key performance indicator among the big dairy groups'. According to Breathnach, a key theme throughout the industry's evolution has been 'the ongoing failure of dairy farmers to exert ownership and control over the co-operatives of which they were members'.[25]

An acquaintance of mine from the early 2000s was a senior executive in a large dairy company and often spoke disparagingly at the time of the involvement of farmers within the organization. As he saw it, the farmers' desire for higher milk prices was fundamentally at odds with the executives' professional objectives, which were to maximize the profitability on which their performance-related bonuses were calculated.

As co-ops merged to form multinational PLCs, the objectives of these corporations grew far beyond their origins in dairying. Kerry Group reached a deal in late 2024 to spin off its fresh milk business to Kerry Co-operative Creameries, in order to focus on being what it calls 'a pure play global taste and nutrition business', with product lines that include nutrition enzymes, proteins, lipids and bases. Glanbia's product portfolio today includes brands such as SlimFast, Optimum Nutrition and Isopure – all a long way from its dairy roots. In 2022, Glanbia PLC sold its shareholding in the dairy co-op called Glanbia Ireland, which rebranded as Tirlán. Tirlán is a fully farmer-owned co-operative, but on a giant scale, with 2,100 employees and sales of around €3 billion.

*

Every five years or so, the Irish government publishes a strategy document for the agri-food sector. 'Food Harvest 2020', as we have seen, was launched in 2010, 'Food Wise 2025' appeared in 2015, and 'Food Vision 2030' was published in 2021.

What all three have in common is that they were conceived, developed and delivered almost exclusively by farming and food industry interests, with limited political oversight and virtually zero environmental input. There was little doubt who was calling the shots in setting the government's agri-food policy. 'Food Wise 2025' was when the then Bord Bia chief executive Tara McCarthy described it as 'industry-owned'.[26]

The committee that produced the most recent national agri-food plan, 'Food Vision 2030', included more than thirty representatives of industry and sectoral interests. Apart from the EPA, the only other representative of the environmental sector – the Environmental Pillar, an umbrella body for twenty-six environmental NGOs – formally resigned from the committee six months before the document was launched. It claimed its input had been disregarded and that the plan had little or nothing to say about vital environmental concerns.[27]

I've spoken to many of the people close to this process, and it is clear that the only reason environmental NGOs were given even a token presence at the 'Food Vision 2030' committee table is that those driving the policy wanted to be able to claim their blessing as a sort of environmentally friendly fig leaf, without actually accepting their input.

Nobody involved in 'Food Vision 2030' can have been in any doubt that Irish agriculture was on a collision course with ecological limits. Committee member Laura Burke of the EPA spelled it out bluntly in the agency's submission to the strategy. Ireland's green reputation was not, she said, supported by the evidence. She noted that economic growth in recent years 'is happening at the expense of the environment, as witnessed by the trends in water quality, emissions and biodiversity all going in the wrong direction'.[28]

Media reaction to the launch of 'Food Vision 2030' struck a markedly sceptical tone. An *Irish Examiner* editorial described as 'disgraceful' the fact that government was accepting recommendations from such an unbalanced group, while noting that the document was 'delivered as if climate collapse was a fantasy in some excited imagination'.[29] A *Sunday Times* editorial cuttingly described the strategy as an attempt to 'spin minimal compliance as leadership'.[30]

*

Ruminant animals, such as cattle and sheep, produce methane via 'enteric fermentation', which is part of their digestive process. The gas is primarily belched out in the form of bovine burps. Methane is an extremely potent greenhouse gas, twenty-eight times more powerful than CO_2 over a hundred-year period, and eighty-four times more potent over a twenty-year timeframe. It accounts for two-thirds of total Irish agricultural emissions.[31] Emissions of methane have increased in line with milk production. Even if there were no other pollution associated with cattle and sheep farming, methane emissions alone, with their catastrophic effects on climate, would be enough to justify a complete rethink of Irish agriculture.

Sadly, methane is far from the only by-product of cattle

and sheep farming. Despite the marketing hype about Ireland's 'natural' advantages in growing grass, the fact remains that farmers are heavily dependent on imported chemical fertilizers to boost grass growth to feed their grazing animals, spending around €1 billion a year on this critical input. The total amount of grass grown in Ireland has doubled since 1960, and this is largely due to the use of chemical fertilizers. The key ingredient in these fertilizers is nitrogen. After fertilizers are applied to pastures, especially on the free-draining soils in the dairy country of the south and east, nitrates leach out into streams and lakes, where they are toxic to fish and cause algae growth that can severely damage animal and plant life. Eutrophication – the overloading of a body of water with nutrients – can also affect the safety of drinking water.

To limit the harmful impacts of nitrogen on water pollution, the EU Nitrates Directive came into effect in 1991, establishing limits on how much nitrogen may be applied per hectare of land.[32] A small number of states, including Ireland, applied for derogations to allow some farms to exceed the upper limit, as long as they undertook certain environmental measures. As the EU Court of Auditors noted in 2021, Ireland is now 'among the highest greenhouse gas emitters per hectare' due to the derogation. This is partly because using more nitrogen means farmers grow more grass, which in turn means more cattle – and thus more methane emissions – on each hectare of land. But there is another way in which nitrogen fertilizer fuels greenhouse emissions. When chemical nitrogen is applied to the land, it leads to the release of nitrous oxide and ammonia, gases that, apart from being dangerous to human health, are also a major source of greenhouse emissions, accounting for a quarter of all GHGs from agriculture in Ireland.[33]

Despite intense political lobbying, the European Commission reduced Ireland's nitrates derogation limit by 12 per cent per hectare from January 2024.[34] The derogation is due to expire in January 2026, but the government intends to apply for yet another renewal. If successful, Ireland is likely to be the only EU country to maintain its derogation, as others have phased it out owing to the extremely harmful effects of nitrogen on water quality.[35]

While chemical fertilizers are imported, there is also a domestically produced source of grass fertilizer: Ireland's national livestock herd produces upwards of 40 million tonnes of slurry a year through its excrement and urine.[36] Like chemical fertilizer, slurry use is regulated by the Nitrates Directive. But even Irish politicians acknowledge that the directive is widely ignored. In October 2023, agriculture minister Charlie McConalogue warned that there was 'too much tolerance' of the illegal spreading of slurry by farmers.[37]

In order to stay within legal nitrates limits, farmers are required to 'export' excess slurry off their farms. In 2022, some 3.6 million cubic metres of cattle slurry was supposedly exported from around 5,900 farms – the equivalent of nearly a third of a million slurry tanker loads, or an unlikely fifty-four loads per farm. What Teagasc has confirmed is that in many cases, the slurry is moving only on paper. A Teagasc nitrates expert stated that stakeholders were aware of this fraudulent behaviour, with many farm advisers being pressured to look the other way.[38]

The best way for intensive dairy farmers to prove that they are indeed in compliance with pollution regulations would for them to be subject to the same EPA pollution licensing as applies to most industries in Ireland, including pig and poultry farms. Any farmer with more than 250 sows

THE LIE OF THE LAND

requires an EPA licence, but thanks to continuous lobbying by the IFA and the Irish Creamery Milk Suppliers Association (ICMSA), a dairy farmer requires no such licensing, regardless of herd size.[39]

Even as water quality has deteriorated, there has been a sharp fall in the number of farm inspections for water pollution being carried out for local authorities. Nationally, there are just eleven full-time equivalent staff to carry out inspections on 135,000 farms.[40] At present, an Irish farm can expect to face a water inspection on average once every hundred years.[41]

There is remarkably little debate in Ireland over the human health impacts of excess nitrogen in our waterways. This is, I suspect, due to lack of public awareness. Denmark is a country of similar population to Ireland, also with a large livestock sector. A 2024 study into the health and economic impact of nitrates found that around 127 annual deaths from colorectal cancer are directly attributable to elevated nitrate levels in drinking water, with an annual cost to the state of around €310 million, in addition to the obvious human toll.[42]

Denmark had been one of only three EU countries with a nitrates derogation. However, the Danish government, mindful of the multiple negative impacts, decided to end its exemption as of July 2024, while the Netherlands is to scrap its derogation by the end of 2025.[43] Politicians and sectoral lobbyists demanding the retention of Ireland's nitrates derogation claim to be defending Irish farming and 'rural Ireland', yet they are also effectively lobbying for more pollution, less biodiversity and more ill health.

The impacts of Ireland's heavy reliance on chemical nitrogen and phosphorus (another fertilizing element) are being most acutely felt in rapidly declining water quality. The EPA reported that nearly half of Ireland's surface waters are in an unsatisfactory condition, while it described declines in water quality in our estuaries and coastal waters as 'alarming'.[44] The trend, the agency noted, is 'getting worse'.[45]

Water pollution issues are also being exacerbated by climate-fuelled torrential downpours, which are now increasingly common in Ireland. A Teagasc expert highlighted an incident in November 2023 when an extreme rainfall event in Wexford led to the equivalent of a year's worth of phosphorus being washed out of a catchment in a twenty-four-hour period.[46]

An EPA study in late 2024 on Lady's Island lake in Co. Wexford, covering 300 hectares, found it to be in an extremely poor ecological condition, probably as a result of 'excessive inputs of nitrogen and phosphorus from agriculture'.[47] Efforts to restore it are expected to cost millions of euros. The level of eutrophication is so severe that, when viewed from a satellite, the lake appears to be glowing blue and green. 'We feel like undertakers to Ireland's natural heritage as we keep writing these obituaries,' said Cillian Roden, a member of the research team.[48]

Politicians were warned in no uncertain terms that this would happen, but the relentless political pressure for agricultural intensification and the promise of an economic boom won the argument. When there is serious money on the table, all other arguments get short shrift.

The dairy intensification reversed years of steady improvements in Ireland's agri emissions, leading instead to an overall increase in emissions from the agriculture sector from 2010 to 2022 of 15 per cent.[49] This is the very timescale within which Ireland's total emissions were supposed to have fallen by at least 20 per cent. Clearly, since our number one source of carbon pollution was actually ramping up instead of being cut back, Ireland has just been playing an elaborate game of charades on our international climate commitments.

*

The one semi-state body that has a full overview of Irish agriculture is Teagasc. It is responsible for agricultural research and development as well as training and advisory services. In 2024 it was in receipt of €168 million in state funding. Where was it during the great liquid gold rush into dairy intensification over the last decade?

We gained a unique insight into the agency's thinking at a conference in Cavan in October 2019. A question from the floor to the panel of speakers asked whether Irish dairying had perhaps expanded too quickly, before the likely consequences had been fully assessed.

In reply, Pat Dillon, Teagasc's director of research, made this astonishing statement:

> Looking back when we were considering what issues would come with 50 per cent expansion, believe it or not we never considered the outcome of the calves in the system – because it was all about how would we get the cow numbers increased, how would we get the milk processed and how would we get enough land for the dairy farmers . . . Fair enough, we probably should have planned.[50]

The only way to produce milk continuously is to impregnate cows annually. The inevitable outcome of this process is the arrival of calves. When you add half a million dairy cows to the herd, you are, by definition, adding a similar number of calves every year.

Dillon's jaw-dropping admission spoke to the frenzied atmosphere that surrounded the lifting of milk quotas in 2015.

The attitude that saw calves as an irrelevant afterthought in the rush to ramp up milk production was painfully exposed in an RTÉ *Prime Time* investigation broadcast in early July 2023, titled 'Milking It: Dairy's Dirty Secret'.[51]

Reporter Fran McNulty's award-winning exposé revealed widespread abuses in the treatment of these vulnerable young animals, especially the live trade, with calves being trucked as far as Spain. How, I wonder, did this chime with the claim in 'Food Harvest 2020' that Ireland operates to 'world class standards' in animal welfare?

Teagasc could hardly claim not to have had ample dairy expertise to call upon. Until recently, five of the eleven members of the Teagasc Authority, including its chair, were dairy farmers, with no representation whatever from tillage, beef farming, organics or the environmental sector.[52] It is somewhat less lopsided today, but there remains zero expert environmental, tillage or organics representation.

The carefully managed messaging from the industry and its boosters has been that any talk of sustained reductions in livestock numbers is off the table. An illustration of what happens when you wander off-script involved Teagasc's own outgoing director, Gerry Boyle, an economist, when in September 2021 he set out the rational case for a shift away from suckler beef production and towards dairy beef.[53] (Suckler herds produce beef cattle, while dairy beef herds are produced from the calves of dairy cows.) This modest proposal drew a firestorm of criticism in the farming press, with agriculture minister Charlie McConalogue weighing in with a promise that suckler beef would always be 'the anchor' of Ireland's beef production system.

The episode shone a light on a divide within the Irish cattle sector. While beef farming operations in Ireland rarely break

even, beef farmers greatly outnumber their dairy counterparts and so comprise a significant part of the overall lobby.
(There are five times more suckler farmers in Ireland than
dairy-only farmers, but Ireland's 1.5 million dairy cattle outnumber suckler cows by nearly two to one.) The narrative that
any form of herd reduction, be it of dairy cows or of beef
cattle, is beyond the pale is enforced by the farming organizations, notably the IFA and ICMSA, with the support of
the farm press. Government proposals for a €2-billion voluntary exit scheme, with compensation for farmers of up to
€3,000 per suckler cow, were shot down in flames by ICOS,
the IFA and ICMSA as well as Meat Industries Ireland and
Dairy Industries Ireland.[54] Teagasc in turn maintains message discipline by developing emissions-reduction scenarios
that explicitly ignore herd reduction or national limits on
milk production as means of reducing emissions. Its various projections all the way out to 2050 envisage a huge dairy
herd and milk production far above today's levels.[55] Teagasc
pays remarkably little attention to the likelihood that serious climate impacts might put the kibosh on its rosy growth
forecasts.

The 'beef levy', a fixed charge on every carcass processed
by meat factories, is also a vital source of funding to the
IFA, accounting for up to half its total revenues and giving
the association a powerful financial interest in maintaining
the size of Ireland's beef herd.[56]

Tensions between the dairy and beef camps have increased
in recent years as the dairy sector has become more politically
assertive. Teagasc estimates that the average Irish dairy farm
earned €89,000 in 2024, while beef farmers averaged just
€9,500 income; and it forecasts average dairy farm income to
hit €113,000 in 2025.[57] The growing economic divide within

Ireland's livestock sector has seen dairy farmers prosper and expand, using their financial muscle to out-compete others, including beef and tillage farmers, to buy and lease land. The dairy boom has been a key driver of the increase in the average price of land in Ireland by 31 per cent between 2020 and 2023.[58]

While dairying is a full-time profession, many Irish beef farmers are part-time, with no realistic prospect of their enterprises ever yielding a living income. The logical option for many would be to exit beef farming altogether, but many beef farmers, especially those who are older, view it as a way of life at least as much as a business. The better alternative for many would be forestry, according to Brendan Kearney, former assistant director of Teagasc; but relatively few are following this path.[59]

*

Ireland, like the EU as a whole, is fundamentally at odds with itself when it comes to agricultural policy. We have binding international commitments to reduce GHG emissions and protect ecosystems. At the same time, we allow – indeed, we actively encourage – farming practices that make it impossible to meet those commitments.

The huge support given by the EU to agriculture via the CAP is heavily skewed in favour of emissions-intensive animal products, a 2024 research paper found, with more than 80 per cent of CAP funding supporting livestock agriculture.[60]

The World Bank recommends that taxpayer support for farming be redirected away from livestock. Such a move, it argued, could lead to overall increases in national income across the EU of 1.6 per cent, while reducing the cost of healthy diets by almost a fifth. This move would slash overall agricultural emissions by 40 per cent across the EU, and proportionately even more in Ireland, given our oversized livestock sector.

Between 2014 and 2020, more than €100 billion in EU agricultural funding was earmarked for tackling climate change. But there has been minimal overall reduction in GHGs from agriculture in the EU since then, and a sharp rise in Irish agri emissions in the same period. A 2021 review by the European Court of Auditors noted that the CAP sets no limit on livestock numbers, nor does it offer incentives to reduce them. It added that CAP market measures include actively promoting animal products.[61]

The Auditors' report also notes that CAP supports climate-unfriendly practices, such as paying farmers who cultivate drained peatlands – a significant issue in Ireland, since drained peatlands are responsible for a disproportionate share of our land-use emissions.

A crucial flaw in EU agricultural policy, according to the European Court of Auditors' report, is that 'EU law does not currently apply a polluter-pays principle to greenhouse gas emissions from agriculture'. This allows highly polluting models of agriculture, specifically meat and dairy production, to avoid paying for their emissions or water pollution impacts.

Three years after the Court of Auditors' report, its findings were endorsed by a report from the European Scientific Advisory Board on Climate Change.[62] It recommended 'emissions pricing' for food in a bid to cut consumption mainly of beef and dairy products in the EU. Critically, the board urged that CAP funding be shifted away from supporting emissions-intensive production, specifically livestock.

An environmental review of Ireland in 2021 by the

Organization for Economic Co-operation and Development (OECD) recommended removing VAT exemptions on fertilizers, animal feeds and agricultural diesel. Removing these subsidies would 'contribute to a more efficient use of resources, improve water quality and reduce greenhouse gases and ammonia emissions', the report added.[63] Its recommendations have been ignored.

The rate at which the climate crisis is impacting European agriculture means that political pressure may grow to force sectors implicated in fuelling it to pay according to their pollution impact. This in turn may have profound effects in the evolution of Irish agricultural policy.

For years, politicians, media commentators and state agencies alike have been gaslighting us about the uniquely clean and green nature of Irish farming. The reality, as we have explored in this chapter, is very different. We've also been told that Ireland's very identity is somehow tied to the agricultural status quo. In truth, that status quo benefits the few – agribusiness PLCs, large-scale ranchers and giant dairy enterprises – at the expense of the many: our struggling tillage, horticulture and organic farmers, as well as the population as a whole. We are paying a heavy price in the form of spiralling greenhouse gas emissions, chronic water pollution and the ongoing obliteration of the natural world in Ireland.

Later in this book, we will delve more deeply into how the myths of Irish agriculture have been conceived and spread. And we will look at how we can reinvent Irish farming and food production in a more nature-friendly and climate-resilient way that will ultimately benefit us all.

3. Powering Our Future

Ireland was once a world leader in renewable electricity. Can history repeat itself?

A century ago, the fledgling Irish state took a monumental gamble when it signed a contract with the German industrial giant Siemens Schuckert to build what would be the world's largest hydro-electric power plant, at Ardnacrusha in Co. Clare. The sums involved were, for the cash-strapped state, astonishing. The amount agreed was £5.2 million, or around a fifth of the state's entire annual budget.

Starting in 1925, some 5,000 Irish and German workers commenced what was by some distance the largest civil engineering project ever undertaken in Ireland. People travelled from all over the country by train on day trips to witness the massive project under construction. In 1927, the Electricity Supply Board (ESB) was set up to manage and co-ordinate national electricity supply. It was not without its critics.[1] One newspaper denounced the move to place energy in state ownership as 'the first fruits of Bolshevism in this country'.[2]

On 22 July 1929, on time and almost exactly on budget, the Taoiseach, W. T. Cosgrave, threw the switch to inaugurate the 86-megawatt Ardnacrusha power station. It was able to produce more than twice as much electricity as Ireland's entire national demand at the time. Franklin D. Roosevelt, then governor of New York, wrote to the ESB that year asking for more information about the project. And in 1933, as US president, he referred to Ardnacrusha when inaugurating the giant Tennessee Valley hydro-electric project.

Ireland was, for a short time, completely self-sufficient in clean electricity. Ardnacrusha's output was so large for the time that the London-based *Morning Post* sniffed that 'the Irish people, with such an excess of power ... may all be electrocuted in their beds'.[3]

Environmental concerns were not, of course, anywhere near top of the agenda of the Free State government, and little consideration was given to the ecological disruption to the River Shannon.[4] A far more pressing issue was the strategic goal of securing national energy independence. It was important then; today, it is even more critical.

The challenge – and opportunity – facing Ireland is the electrification of our entire economy and society, including sectors such as transport, industry, energy production and home heating that are heavily dependent on fossil fuels. A transition on this scale is massively challenging. But with sustained political will, joined-up thinking at the regulatory level, technical expertise and a combination of state and private funding, we can once again pull off a project that will help to prepare Ireland for the dangerous and uncertain decades ahead.

*

Despite significant growth in renewable energy since the turn of the century, some 85 per cent of our total energy needs across transport, heating and power production come from fossil fuels, a high level of dependency by EU and international standards; and around three-fifths of our total greenhouse emissions are energy-related.[5] Ireland spends well over €1 million an hour, twenty-four hours a day, on imported fossil fuels.[6]

If Ardnacrusha was the first great renewable-energy

project in independent Ireland, it took seventy-five years for the second to come along. In 2004, Airtricity commenced operation of the Arklow Bank Wind Park, Ireland's first off-shore array and the world's first installation of wind turbines greater than 3MW. A total of seven turbines were installed in phase one, producing a combined 25MW, or enough electricity for around 20,000 homes.

That was supposed to be just the start. Phase two was to have added another 193 turbines in a huge array generating up to 520MW, or similar to a gas-fired power station. This project was cancelled in 2007, largely due to a government decision to support only onshore wind projects, as offshore was deemed too expensive and risky. Following the 2008 global financial crisis, capital-intensive offshore wind disappeared from the national radar for almost a decade.

This was a major missed opportunity, and we are now playing catch-up. But it is important to acknowledge where we have been successful, and that is in renewable electricity from onshore wind energy. Twenty years ago, a mere 4 per cent of Ireland's electricity was produced from renewables. By 2023, this had risen ten-fold to just over 40 per cent, the overwhelming bulk of which was from onshore wind farms.

Should Ireland be regarded as a world leader in renewable energy, or as a chronic fossil-fuel laggard? It may be possible to make both arguments simultaneously. Ireland ranks in the top five globally for installed onshore wind energy capacity per person and for the contribution made by wind energy to meeting national electricity demand. Apart from the modest Arklow Bank array, the vast bulk of the estimated 5 gigawatts of installed wind energy in Ireland comes from just over 300 wind farms dotted across the country. In an average year, Ireland's wind farms avoid the release of around 4.5 million tonnes of CO_2 emissions from other energy sources. This is similar to total emissions from all of Ireland's households.[7]

With low targets and even lower expectations attached to Ireland's agricultural emissions, much of the heavy lifting in terms of meeting our emissions targets will have to be done by the energy sector. The Climate Action Plan aims to cut emissions from electricity production by 75 per cent versus 2018 levels, reducing them to around 3 million tonnes of CO2 by 2030. However, this target is greatly complicated by the fact that in the same period total electricity demand is expected to increase by up to 50 per cent, with data centres the most voracious new consumers of electricity. The plan envisages doubling our onshore wind production by the end of this decade to 9GW, as well as increasing our offshore wind from its current 25MW to 7GW, representing a 280-fold increase. But it is reckoned that only six offshore wind farms, with a combined capacity of 4.3GW, have a realistic chance of being deployed by 2030.[8]

Even more dramatic progress is needed from solar power. Starting from virtually zero at the beginning of this decade, we are aiming to have a whopping 8GW installed by 2030. This is roughly the equivalent of installing more than 15 million individual solar panels. This may not be as far-fetched as it sounds. As we'll explore later in this chapter, worldwide, the solar revolution continues to vastly exceed even the most optimistic projections, and rapid deployment is far less technically challenging than building offshore wind facilities. Despite our frequent grey skies, Ireland's potential in this area is much greater than generally assumed.

Ireland has a massive opportunity in offshore wind. Our maritime area is seven times larger than our land mass, and our wind speeds are among the best in the world. Higher wind speeds at sea mean wind turbine efficiency can be double that of an equivalent land-based turbine.[9] Over the next quarter-century, up to 37GW of offshore wind energy could be brought online in Irish waters, with a likely value of at least €38 billion to the Irish economy, according to industry estimates.[10] The state's own policy on offshore wind makes similarly ambitious projections. The amount of wind energy foreseen in these projections is at least five times greater than Ireland's entire electricity demand at present.[11]

In mid-2023, Ireland auctioned the rights to develop four major offshore projects, with a total capacity of 3GW.[12] This represents a quarter of what is expected to be our total national electricity demand by 2030. The largest of the four projects is the Codling Wind Park, comprising between sixty and seventy-five large turbines located between thirteen and twenty-two kilometres off the Wicklow coast.[13] This one wind farm will produce fifteen times more electricity than the Ardnacrusha hydro plant. A single rotation of one of these giant turbine blades produces enough electricity to power a typical Irish home for two days.

The average price secured at this auction was €86/MWh, which translates into a wholesale price of 8.6 cents per unit of electricity – which compares very favourably with the average wholesale price of electricity in Ireland in early 2025.[14] More wind energy underpins cheaper electricity prices. In time, this and other large-scale offshore wind projects will save Irish consumers hundreds of millions of euros annually. Lower prices are good, but *stable* prices are equally important. The more renewable electricity capacity that Ireland has, the less dependent we are on imported fossil fuels, whose prices are unpredictable and whose availability may be affected by events thousands of kilometres away – as we were forcefully reminded following Russia's invasion of Ukraine. Global geopolitical instability has increased alarmingly since the re-election in the US of Donald Trump, who immediately set about upending trade and security alliances. It is critical that Ireland take control of its energy future and break our dependence on foreign fossil fuels.

All new infrastructure, whether on land or at sea, requires planning approval. Given that a single wind-farm application can be up to 9,000 pages long, this is a hugely time-consuming process, and Ireland's under-resourced planning system has struggled to cope, leading to significant delays. The need for a radically different approach was underlined in November 2021, when Norwegian energy giant Equinor announced it was pulling out of a proposed €2-billion joint venture with the ESB to develop a wind farm near Moneypoint in Co. Clare, citing its frustration at ongoing delays in the planning process.[15]

Paul Deane of University College Cork (UCC) has argued persuasively that the planning process for renewable-energy projects needs to be put on an 'emergency' footing for fast-tracking.[16] A significant step in the right direction occurred in July 2023 with the launch of MARA, Ireland's Marine Area Regulatory Authority. MARA is chaired by Vice-Admiral Mark Mellett, former chief of staff of Ireland's defence forces, who also holds a PhD in environmental governance. Its purpose is to streamline and simplify what has been often a chaotic and disjointed process facing those trying to develop offshore renewable-energy projects.

While land-based wind farms have proliferated in Ireland, they have encountered significant resistance along the way, some of it from unexpected sources. Some years ago I was cornered at a reception by a well-known environmentalist

who was aghast that wind turbines had, in her view, ruined
the pristine rural landscape where she lived. My attempts at
pointing out that wind farms meant clean, home-produced
electricity cut no ice. Of course, whether you see wind turbines as a visual blight on the landscape or as elegant symbols
of sustainability is entirely subjective, but surely, in the teeth
of a climate emergency, we must be past the point where
any of us, environmentalists included, can place our own
aesthetic preferences ahead of the urgent necessity to decarbonize our energy systems.

Climate action generally, and renewable energy specifically,
received a welcome boost in early 2025 in a case taken by the
Norwegian hydropower giant Statkraft against the refusal by
An Bord Pleanála (ABP) to grant it permission to build a wind
farm at Coolglass in Co. Laois. In his judgment, Mr Justice
Richard Humphreys was scathing of ABP's failure to take
account of Section 15 of the Climate Action and Low Carbon
Development Act, which directs public bodies to take climate impacts into account in their decision-making. The ABP
inspector had given far greater priority to visual impacts than
to Ireland's compliance with binding national and EU climate
targets, the judge said. He added that only properly trained
people 'who understand climate issues at a deep level should
be allowed near projects to which those issues are relevant'.[17]

This High Court ruling is likely to be very significant.
Upcoming road-building projects, so beloved of local politicians, may well fall foul of this landmark decision, unless of
course it is overturned on appeal.

In mid-2023 I attended a public information meeting in
Dún Laoghaire on the proposed RWE Dublin Array, an
824MW offshore wind farm, for which formal planning
application was lodged in early 2025.[18] While the overall

tenor of the discussion was constructive, the most sustained opposition – apart from locals who simply didn't like the idea of being able to see wind turbines on the horizon – came from people expressing concern about the ecological impacts of installing and operating offshore wind farms.

Every form of energy production, including renewables, has its downsides. The manufacturing of solar panels and wind turbines involves energy, water use and raw materials, some of which are toxic. Wind farms located on upland bogs have sometimes triggered landslides. In December 2016, a worker was killed as a result of a landslide at the Derrysallagh Windfarm in Co. Roscommon. Offshore wind farms, too, can have negative impacts. Fish and sea mammal behaviour and migration can be disturbed, especially during the construction phase. The overall impact of wind farms on bird populations is negative as well, with adverse effects on breeding, migration and survival rates.

Legitimate environmental concerns, when supported by evidence, should of course be taken seriously, and I have no doubt that most of the people objecting are acting with what they see as the best of intentions. But what I did not hear that evening, from any of the people opposing offshore wind energy, was what they would suggest as an alternative.

We are in a dire climate emergency that affects the oceans as much as the land. Along with overfishing, bottom trawling (which, apart from destroying marine habitats, is also a significant source of CO_2 from disturbed sea floor sediments) and pollution, the greatest threats to marine biodiversity are rising sea temperatures and ocean acidification. When oil tankers sink or offshore oil rigs leak or explode, the result is ecological devastation. Offshore oil and gas exploration involves massive seismic blasting to map the sea floor. The noise and shock waves from this exploration are a major threat to marine life, especially mammals.[19] Further, around 40 per cent of the world's shipping at any given time is hauling gas, oil or coal, which threatens marine life with noise, pollution and collisions in the process. The evidence of the threat to our seas from the continued burning of fossil fuels is overwhelmingly greater than the risks associated with offshore renewable-energy systems.

While Ireland hums and haws over going all-in on renewables, it's instructive to see how Denmark, a country of similar population, has fared. It was a pioneer in wind energy in the 1970s and over the decades has built up a substantial industry employing more than 33,000 people with a turnover of €19 billion – similar in size to Ireland's entire agriculture and food sector.[20] The world's largest wind turbine manufacturer, Vestas, is Danish, and it is a major supplier of wind turbines to Ireland.[21] Well over half of Denmark's total electricity production is now supplied by wind energy. Its national target by 2030 is to have 55 per cent of all its energy from renewables.[22]

Even in a country that has fully embraced offshore wind, however, the transition is not always plain sailing. In December 2024, an auction of 3GW of offshore wind in the Danish North Sea ended in failure, with no bids received.[23] Industry sources blamed rising material and construction costs for turbines and uncertainty over the prices available for renewable electricity.

Familiarity seems to have bred contentment as the Danish public has embraced offshore wind turbines as symbols of the clean energy transition. Every year, thousands of people take boat trips that combine viewing seal colonies with visits to offshore wind farms.[24] An ecological benefit of offshore wind farms is that, since fishing is restricted, these become de facto marine sanctuaries, with the bases of the turbines often forming artificial reefs. Marine areas set aside for wind farms will be spared from bottom trawling and are likely to evolve into new marine nurseries, which in time will also benefit fisheries, despite the short-term protestations of the fishing industry.[25]

In 2023 Ireland was confirmed as one of a group of eight European countries that have pledged to quadruple wind energy by 2030, to develop artificial 'islands' connecting off-shore wind farms and to work to standardize technologies to help scale up production.[26] The consortium of nations aim to ratchet up combined offshore wind production from today's 30GW to 120GW by 2030 and 300GW by 2050.

Taoiseach Leo Varadkar said that, for its part, Ireland aimed to build enough offshore wind farms to produce five times Ireland's entire current electricity demand and to become a major energy exporter. Upgrading Ireland's grid to cope with this massive increase in electricity would, Varad-kar added, cost billions, but despite the outlay, the economic benefit to Ireland would be orders of magnitude greater in the decades ahead.[27]

Almost twelve months later, the enterprise minister, Simon Coveney, launched the state's 'Offshore Wind Indus-trial Strategy', a reversal of roles that saw him working to solve the climate crisis, rather than contribute to it as he had when agriculture minister. The guiding ambition is to deliver at least 37GW of offshore wind capacity by 2050.[28] There are, however, formidable technical and logistical challenges to be overcome before this ambition can even be addressed. Delays and uncertainty in the planning process create a ripple effect, as Ireland is just one of a number of countries com-peting for finite resources.

Globally, there are no more than ten ships capable of handling the 15MW offshore turbines and bringing them into position. These need to be booked years in advance, as do the turbines themselves, and the laying of the extensive cabling needed to bring power back to shore, as Noel Cunniffe, CEO of Wind Energy Ireland, explained to me. At present, no port in the Republic of Ireland has the infrastructure to accommodate components for massive offshore wind turbines. Belfast has the only port on the island with the required capacity.

'We've been asking for several years for state funding to be put into Cork, Rosslare, and Shannon Foynes to try to get these ports ready to be able to construct these projects,' Cunniffe said. If Ireland fails to build this critical port infrastructure, we are likely to see turbines destined for Irish waters being assembled as far afield as France or even Norway, with a huge loss in terms of skilled work and value-added for Ireland. The likely cost of upgrading each harbour would be of the order of €100 to €300 million, which is relatively small beer when we consider that Ireland's capacity for offshore projects is valued in the tens of billions. There is, however, another catch. The current national ports policy prevents the state from funding ports.[29] This situation needs to change as a matter of urgency, Cunniffe argued.

Another potentially significant source of offshore energy comes from below the surface of the ocean. Ireland's Centre for Ocean Energy Research (COER) at Maynooth University argues that wave power could be a complementary technology to offshore wind.[30] In theory, wave energy could meet half of Ireland's entire electricity needs, but its implementation is technically challenging.[31] In September 2024, the EU approved a research project to advance commercialization of wave power; this brings together partners from

Ireland and thirteen other countries to explore its viability. Wave power may well mature as a technology, but it's unlikely to be a significant factor for at least the next decade or two.

*

While the rollout of renewables onto the Irish grid has been a relatively bright spot amidst Ireland's ongoing heavy dependence on imported fossil fuels, the country's fast-growing infatuation with data centres is pulling us in the wrong direction. The share of Ireland's total electricity production used by data centres rose from 5 per cent in 2015 to 21 per cent just eight years later. By 2023, some eighty-two data centres were using far more electricity than all of Ireland's urban households combined.[32] In response to concerns about the rapid growth of data centres, EirGrid in 2021 published new rules that effectively put an embargo on grid connections for new centres in the greater Dublin area until 2028.[33]

The government has stated that data centres have contributed over €7 billion to the economy since 2010.[34] Our dependence on the presence of digital giants including Apple, Microsoft, Facebook, Google and Amazon as major employers and corporation-tax payers has made the Irish state leery of rejecting their demands for ever more data centres to be constructed.

The digital behemoths are voracious energy consumers. Google announced in July 2024 that its global carbon emissions had increased by almost 50 per cent in just the last five years, to 14.3 million tonnes. Much of the additional energy demand is being created by the intense computing associated with artificial intelligence. With AI functionality now being embedded into millions of devices worldwide, energy demand will continue to spiral for the foreseeable future. For Ireland, trying to decarbonize our electricity system by adding in renewable energy while also allowing ever more data centres to be built here means we are essentially running to stand still.

*

Shortly before his death in 1931, electricity pioneer Thomas Edison reportedly confided to his friends Henry Ford and Harvey Firestone: 'I'd put my money on the sun and solar energy. What a source of power! I hope we don't have to wait until oil and coal run out before we tackle that.'[35]

Solar energy has been growing globally at a phenomenal rate, and the cost per megawatt of solar panels has plummeted. The International Energy Agency's *World Energy Outlook 2020* described solar photovoltaics (PV) as 'now the cheapest source of electricity in history'. Globally, in 2004, it took a full year to install 1 GW of solar power. By 2010, it took less than a month. By 2024, the world was adding more than 1 GW of solar power every day.[36]

Ireland is hardly famed for its sunshine, and this has fuelled scepticism about the role that solar PV can play in meeting our energy needs. When I interviewed Sarah Mc Cormack, professor of environmental engineering at Trinity College Dublin, about solar energy in 2021, she told me it had been difficult to attract much interest – or funding – to solar projects in Ireland. 'The general attitude we encountered is that renewable energy is all about wind and ocean power in Ireland; the mindset was that we're not sunny enough for solar power.' In reality, solar PV does not require direct sunshine. It requires only daylight, and Ireland has that in abundance.

Since I spoke to Mc Cormack, Ireland's solar sector has grown dramatically. By mid-2024, the peak solar capacity on the electricity grid was comparable to the output of three gas-fired power stations.[37] Around half of this was coming from industrial and agricultural solar PV plants, but a third was being supplied by microgeneration, mostly from rooftop solar on over 100,000 households around the country that had installed their own solar panels, while ESB Networks is processing around 750 new applications every week.[38] At peak, solar supplies around one-sixth of Ireland's electricity.[39]

This should be only the beginning. Solar can be built and deployed far more quickly than almost any other energy system. Apart from when located on rooftops, it requires land. The Climate Action Plan envisages 5 GW of solar PV on the grid in 2025, rising to 8 GW by 2030. This amount of solar would require a mere 0.2 per cent of our available farmland.[40] This presents an opportunity to farmers to enter long-term leasing agreements for some of their land for solar sites, achieving much higher income per hectare than is possible when using the same land for cattle or sheep grazing.[41] Because solar panels are spaced out and typically cover only 20 to 40 per cent of the land within the solar farm, there remains scope to graze smaller animals like sheep, or to grow wildflowers under the panels, and farmers' direct farm payments can still be claimed if sheep are retained. Leasing is by far the most popular approach taken by solar farm developers, with typical leases running from fifteen to twenty-five years, broadly matching the duration of planning permissions for solar farms.

A study of solar farms in the UK indicates they can offer another huge, if unexpected, benefit: biodiversity protection. At most farms that converted to solar, biodiversity was found to be improved. At the end of a leasing period, a farmer may choose to return the land to other uses. And if the revenue allows livestock farmers to boost their income while carrying fewer animals, there will be a further environmental benefit in reduced methane and other pollutants, and enhanced bio-diversity.[42] Solar arrays also offset more carbon emissions every year than even the equivalent land area planted with trees.[43]

Despite all the positives, solar farms in Ireland often face intense local opposition. For example, in early 2025 residents in East Cork rallied to oppose a proposed solar farm at Green-hills, which is currently a large dairy farm. The solar operation will produce enough clean electricity to power over 50,000 homes. Local TD and dairy farmer James O'Connor spoke against the proposed development in the Dáil, warning of 'seriously devastating consequences for the dairy industry'.[44]

This fear is unfounded – as we've seen, achieving the government's solar energy targets will require a mere 0.25 per cent of Ireland's agricultural land area – and the Greenhills project, if it goes ahead, can only be viewed as good news. A switch from dairy to solar is probably the single most climate-positive transition a farm can make.

Most of us don't own large dairy farms. But if you have a rooftop, you too can be an electricity producer as well as consumer. The removal of VAT on solar panels as well as scrapping the need for planning permission in 2022 helped to cut the installation cost for the average householder by around €1,000 and made the process more straightfor-ward. The price for a ten-panel system suitable for a typical household is around €8,400, less an €1,800 grant from the Sustainable Energy Authority of Ireland (SEAI), leaving a net cost of around €6,600.[45] While an outlay like that may be a no-brainer for many middle-class families, it is simply too expensive for most people with limited income or savings. What is urgently needed is a low- or zero-interest scheme to allow people to pay a small amount upfront and have the balance paid off their electricity bill over several years. This would be mostly self-financing thanks to the annual savings of over €1,000 year on solar electricity, and it would ease the sting of the big upfront outlay that puts solar PV beyond the reach of many.

The government's Microgeneration Support Scheme allows households to sell any excess electricity back to the grid, usually receiving around 20 cents per unit.[46] All these changes mean a householder can expect to have received payback on their solar panels within around six years. If you have an electric vehicle, you can charge up during daylight hours, meaning it's quite possible you could cut your motoring fuel costs to close to zero. And by the way, solar panels will cut the amount of carbon emitted from an average home by 1.3 tonnes a year.

You don't have to be green for this to make eminent sense. Ryanair boss Michael O'Leary, who has long been dismissive of concerns about climate change, admits to being 'astonished' at the impact of installing a solar and battery array at his farm near Mullingar. 'We're here in north Westmeath, it's not the Costa del Sol, and yet we are generating remarkable amounts of power,' he noted. O'Leary's 90kW system, supported by generous government grants for agri-solar, saves him around €36,000 a year in electricity costs and he expects it to be fully paid for in just over five years.[47]

In 2022, the French government brought in a mandate requiring all car parks with spaces for eighty or more vehicles to be covered by solar panels.[48] This simple yet brilliant measure, which has the added benefit of providing shade for parked cars, is expected to yield 11GW of power, which is the peak daytime equivalent of the output of a dozen nuclear plants. Ireland's sprawling outdoor car parks should all be subject to a similar mandate. Wicklow County Council has already installed such an array over its car park; it is the largest such array in Ireland, and now produces 40 per cent of the total electrical baseload demand for the council's headquarters.[49]

The government's decision to roll out a €50-million school solar PV programme may come to be seen as an extremely smart investment. Apart from fully funding up to sixteen solar panels per school, it involves a whole generation of schoolchildren in learning at first hand about renewable electricity and seeing it power their own schools in real time. Many will no doubt be pestering their parents to install solar panels at home, while others may well be inspired to consider a career in renewable energy.

*

Thanks to the increased penetration of wind and solar on the grid and the expanded availability of battery storage, in 2024 EirGrid cut the number of large fossil-fuel power stations required to be operational at any given time from five to four. The maximum amount of renewables the Irish grid can support is, at the time of writing, about 75 per cent.

The critical challenge in increasing that figure is to be able to store sufficient energy to meet our needs when the wind isn't blowing and there is no daylight. Until just a few years ago, the only form of electricity storage on the Irish grid was the pumped storage system at Turlough Hill in Co. Wicklow, which became operational in 1974. This facility uses surplus electricity at times of low demand to pump water uphill to a large reservoir. During surges in electricity demand, the turbines at Turlough Hill can go from standstill to full 270MW generation in seventy seconds, and provide power for up to six hours.[50] Pumped hydro storage facilities are, however, costly to plan and build and depend entirely on suitable geography, which may explain why none have been added in the half-century since Turlough Hill came online. A second pumped hydro facility at Silvermines in Co. Tipperary is currently in the design and assessment phase, but realistically this technology will meet only a small part of our energy storage needs.

Battery technology has rapidly matured in recent years, with costs falling fast. To date, around 1GW of battery storage has been added to the system. These facilities store surplus grid energy and can provide near-instantaneous power when required.

Another element involved in stabilizing the Irish electrical grid is interconnection.[51] Interconnectors are high-voltage land and undersea cables that connect to the electrical grids of neighbouring countries. They allow instantaneous sharing of electricity. For Ireland, multiple interconnectors to the UK and the continent will allow us to export the hoped-for electricity surpluses produced by offshore wind farms. An integrated and interconnected European 'supergrid' allows more and more renewable energy to be brought online and to be instantly directed to where demand is highest. The wind is almost always blowing somewhere, so interconnection helps smooth out the variability associated with renewable power. Ireland now has two 500MW undersea interconnectors to Britain – the latest, known as Greenlink, went live in January 2025 – and we also have access to the Moyle Interconnector, which links Northern Ireland to Scotland.[52] (Ireland and Northern Ireland have operated as an integrated single electricity market since 2007.) Imported electricity accounted for about a tenth of total usage in Ireland in 2024. The 'Celtic Interconnector' to France is due for completion in 2027.

No system is bulletproof. Early in the morning of 14 May

2024, both interconnections to the UK suddenly failed within milliseconds of one another: one failure almost certainly triggered the other. Battery storage facilities kicked in instantly to plug the energy gap for the short period before an additional gas-fired power plant came online to cover the shortfall. At the time they failed, the Interconnectors were supplying around a fifth of Ireland's (off-peak) electrical consumption. As Paul Deane explained, 'Interconnectors are great at sharing solutions, but they're also very good at sharing problems, so if there's a problem in the French grid or UK grid, those problems can ripple their way over to the Irish system, either in the form of prices, or of physical interruption.' While recognizing the value of interconnectors, Deane believes that in the longer term we should not be depending on them to provide more than a tenth of our electricity needs at any given time.

While battery storage is an efficient means of buffering our grid, its key limitation is that it is suitable only for short periods. For instance, Ireland's largest battery storage facility, located at the ESB's Poolbeg site in Dublin, can provide backup electricity for two hours using a lithium-ion battery array that cost €300 million.[53] There are, however, some very promising new long-duration options now being developed, including iron–air batteries. A planning application has been lodged to build Europe's first large-scale iron–air battery facility near Buncrana, Co. Donegal. The facility would be capable of supplying 10MW of power continuously for a hundred hours.[54]

The nightmare scenario for electricity grid operators is what is known as the *Dunkelflaute*, a German term meaning 'dark doldrums'. This is an extended anticyclonic period of low winds, usually occurring during the winter or early spring, when solar production would also be very low. Interconnectors can probably provide around a quarter of our electricity needs, but unless we can dramatically scale up clean energy backups and long-duration energy storage systems, it will be extremely difficult to fully decarbonize the grid, especially as electricity demand continues to increase.

The challenge of harnessing wind energy via mass storage systems is very much on the national agenda. The ESB is working with Bord Gáis Energy and dCarbonX on a plan, known as the Kestrel Project, to redevelop the decommissioned Kinsale offshore gas fields for renewable gas and green hydrogen (hydrogen produced from renewable energy).[55]

Large-scale offshore wind turbine arrays routinely generate far more electricity than can be consumed on the grid. Using the spare electricity from either wind or solar arrays for electrolysis, water can be converted into hydrogen gas, which is then either used to fuel gas-fired power stations, or compressed and pumped into a mass storage facility where it can be kept on standby in the event of a *Dunkelflaute*.

Hydrogen is highly unlikely to have any meaningful role in directly fuelling vehicles or in warming homes. The fossil-fuel industry is interested in hydrogen being manufactured using natural gas, but from a carbon pollution point of view, this is a cul-de-sac whose only possible purpose is to allow these companies to continue selling their climate-destroying products by greenwashing fossil gas as 'clean' hydrogen.[56] The involvement of Bord Gáis Energy in the Kestrel Project may be a red flag: despite its advertising campaigns alluding to 'renewable gas', more than 99 per cent of what the company sells is fossil gas. This will, it claims, change radically in the near future, with 85 per cent of the company's power being 'sourced green' in the next decade, the balance coming from fossil gas.[57] These promises should be treated with some scepticism. It is questionable whether Ireland's gas network has any meaningful role whatever to play in a zero-carbon electrified future, with the grid facing a likely death spiral of declining use and rising costs in the next decade.[58]

The government aims to replace around a tenth of Ireland's fossil-gas usage with biomethane by 2030. This form of gas is made by a process called anaerobic digestion (AD), in which organic material including grass, slurry and food waste is converted first into biogas and then purified into biomethane. According to Teagasc, in order to feed silage into AD plants, some 120,000 hectares of grassland would be needed, as well as 20 per cent of all the winter cattle slurry produced in Ireland.[59]

Originally, development of AD was intended to give farmers an additional stream of income, which would allow them to reduce livestock numbers, which would in turn make them less vulnerable to the fodder shortages that have bedevilled Ireland in recent years as the result of extreme weather events. However, Teagasc now appears to be implying that, by increasing grassland productivity, an AD industry can be supported 'with little livestock displacement', according to its director of research, Pat Dillon.

This seems an optimistic reading of the evidence. Fodder crises like the severe one in 2018 are twice as likely to recur in the coming decades as a result of climate change, according to the EPA-funded ClimAg research project.[60] In the summer of 2024, two in three Irish dairy farmers were reportedly considering cutting stock numbers because of fodder shortages following an extremely wet spring that badly affected grass growth.[61] Diverting large amounts of grass to produce biomethane without a parallel commitment to reduce live-stock numbers seems like a recipe for disaster.

The government is planning to provide an initial €40 million in funding to support the development of ten anaerobic digestion plants. The overall aim is to have 200 such plants dotted around the country, which would suggest additional government support approaching €800 million will be needed. Each of these plants would have to be fed by slurry and silage ferried by fleets of trucks, and these plants would in turn have to be operated to the highest industrial standards, including being fitted with complex and expensive monitoring systems capable of automatically shutting them down in the event of methane leakage. The gas would in turn have to be collected by still more trucks on the rural road network and taken to central processing facilities. Any leakage of methane would have highly negative climate impacts. The economic argument is also unconvincing. A US study found that to extract $68 of gas from cow's manure cost $294, 'starkly illustrating how dependent the entire renewable gas enterprise is on government subsidies'.[62] A hectare of land covered in solar panels will over a year produce around a hundred times more usable energy than the same hectare dedicated to bioenergy.[63]

*

Ireland's transition to an electrified future will reduce our dependence on costly imported fossil fuels in an increasingly volatile geopolitical landscape. But as recent events have shown, this transition is itself threatened by the extreme weather events that are fuelled by climate change. For instance, in December 2024 Storm Darragh knocked out power supplies to nearly 400,000 households. Less than four weeks later, Storm Éowyn, among the worst storms ever to hit Ireland, left 768,000 households without electricity – in some cases for over a fortnight – while also disabling water supplies to more than 200,000 households. Telecommunications was also badly hit. Flattened Sitka spruce plantations pulled down electricity wires in many areas.[64]

How can we mitigate the effects of future storms on domestic energy supply? While mass battery storage provides grid resilience at a national level, battery packs could similarly provide resilience for individual homes or to power micro-grids to keep the lights on for isolated communities. An upcycled EV battery could maintain electrical supply to a typical home for three or four days, and even longer if connected to solar PV. Plummeting prices for battery storage mean this is now feasible for many households, but generous grants should also be made available to help lower-income families, especially in rural areas.

ESB Networks is already investing billions of euros to help weatherize our grid, including replacing millions of wooden poles with more durable materials. To cope with climate change, our utilities need to be resilient enough to withstand wind speeds of 200kph, according to Peter Thorne, chair of the CCAC's Adaptation Committee. As he told the *Irish Times* in the immediate aftermath of Storm Éowyn, 'In the "electrification of everything" – heating, cooking, transport – we can't afford to have days, weeks, offline.'[65]

*

There is one more important piece in the energy puzzle. To meet the drastic decarbonization targets needed to avoid climate breakdown, we have to reduce the *demand* for energy. 'We don't have enough time to simply "technology" our way out of this crisis,' Hannah Daly, professor in sustainable energy at UCC, told me. Much of the energy use in Ireland is wasteful, including, as we shall explore shortly, energy-inefficient housing, and spatial planning that forces people into costly car-dependence while doing little to enhance people's happiness or health.

One of the biggest challenges in energy systems is the Jevons paradox, or energy rebound, whereby increases in efficiency lead to more consumption, which in turn leads to higher overall energy use and carbon emissions.[66] For example, introducing a more efficient jet engine might sound 'green', but if it allows an airline to lower the ticket price, thus leading to more flights, the net effect is ever more emissions.

This was addressed head-on by SEAI chief executive William Walsh in early 2025. Technology alone will not suffice, he pointed out. 'We must also embrace measures that seek to meet our societal needs at a reduced rate of consumption.' This requires, he added, 'building a policy environment that focuses on sufficiency and avoids overconsumption'.[67]

This is exactly the kind of language we need to deploy, yet you will find nothing remotely approaching this sentiment in the 2025 Programme for Government, a document that is built on business-as-usual assumptions and that carefully avoids being honest with the Irish public regarding the limits of consumption.

A clear opportunity remains for Ireland to achieve the key strategic goal of energy independence while becoming a world leader in wind energy and developing a major new export sector in clean electricity. But this is by no means inevitable, and political commitment is essential. Moves by the state to approve a €300-million liquefied natural gas terminal in the Shannon estuary represent a backward step.[68] It is crucial that we do not lose sight of the overwhelming urgency of ending our dependence on imported dirty energy from some of the world's most unpredictable and indeed hostile sources.

64

4. Ireland on the Move

I have crossed Dún Laoghaire's main street, Upper George's Street, many hundreds of times since our business moved to the town in the mid-1990s. But it was only around a decade later, when my daughters were starting pre-school, that I really noticed some glaring flaws in the design of the town. By then I was living locally and was able to walk to work. The Montessori happened to be on the same street as my office, so it became a quotidian pleasure to take the two girls by the hand and walk them to and from school.

That's where things got strange. Our route took us precisely halfway down Upper George's Street, with no traffic lights or pedestrian crossing whatever as a constant stream of traffic snaked by in both directions. On many mornings, my only option was to pick my daughters up, one under each arm, wait for a gap in the traffic, and dash across the street. Detouring to the nearest traffic light would have involved at least doubling the length of our morning walk.

The situation on Upper George's Street, I soon came to realize, was not an anomaly. The DART station and bus terminus in Dún Laoghaire caters for hundreds of people every hour, yet to cross the busy road to access the station requires waiting several minutes for the pedestrian crossing light. When the light does finally change, it remains green for . . . precisely six seconds. It's another small but telling daily reminder that public transport users and pedestrians are the poor relations compared with the overarching imperative of keeping road traffic moving.

In the case of Upper George's Street, I did some measurements and some rough sketches, then contacted a local councillor, who agreed the lack of a safe way to cross a busy street in the heart of the town was an absurd oversight. She pursued the matter at council level. Around three years later, a set of pedestrian traffic lights was finally installed at the point where George's Street intersects with foot traffic from the town library and the local hotel.

By then, my girls had moved to a school a couple of kilometres outside the town. It was situated on a wide roadway with no bike lanes and no safe way of crossing, even though the school catered to hundreds of children. Unsurprisingly, every morning and afternoon saw streams of cars, ours included, dropping and collecting students. Almost nobody cycled and there wasn't even a dedicated bike rack in the school, such was the lack of demand.

If you were to set about designing a system that baked in high emissions, congestion and car dependency from the earliest age, it would, I suspect, look a lot like this.

*

In the decades since 1990, emissions from Ireland's transport sector have risen by around 130 per cent.[1] Transport is, after agriculture, our second largest source of greenhouse-gas pollution, accounting for just over a fifth of the total.

Ireland now has the second highest car dependency in the EU.[2] The overwhelming dominance of the private car in Irish transport is often presented as both natural and inevitable. In reality, it is neither.

Before the first car appeared in Ireland, Dublin had a thriving public transport system. In the mid-1890s, Dublin's first electrified tram was inaugurated, running from Dolly-mount to Nelson's Pillar on what is now O'Connell Street. By 1904, Dublin's tram system was regarded as among the most extensive and impressive in the world, with transport experts from other cities coming to see it at first hand.

To electrify the line from the city to Dalkey, the Dublin United Tramways Company built its own power station on Shelbourne Road that was capable of simultaneously power-ing fifty trams. A 1907 guide to Dublin described the city's public transport as 'second to no city in Europe'.[3]

Beyond the capital city, in 1920 Ireland boasted over 5,600 kilometres of railways; but by the 1950s and 1960s more than half the network had disappeared. As a small child, I was brought by my parents to witness the dismantling of the rail-way bridge in Kilkenny that followed the shutdown of the rail link to Castlecomer some years earlier. Today, there are less than 1,700 kilometres of rail lines. Meanwhile, Ireland's road density of 20.2 kilometres per 1,000 people is more than twice the EU average and more than three times that of the UK.[4] Maintaining this extensive road network imposes huge costs on local authorities and the central exchequer, espe-cially as cars have become on average 400 kilograms heavier in recent years.[5]

In the decades up to 2020, state transport spending was strongly skewed towards roads, which took up around two-thirds of investment, compared with just a third for public transport, while active transport barely featured at all. Such an approach is, of course, self-perpetuating. With invest-ment flooding into road building and upgrading while public transport languished, it is hardly surprising that the Irish public took to their cars in ever increasing numbers.

As private cars proliferate, they undermine the financial

viability of public transport, sending it into a death spiral whereby services are reduced, leading to a loss of public confidence, leading to more and more people buying cars, and so on. This in turn deters governments from investing in public transport. The result, in Ireland, is that the car is king, accounting for two-thirds of all journeys taken.[6] Almost half of all car journeys taken are for distances of five kilometres or less. It is true that we have low population density, but this alone cannot explain these remarkable statistics. Three out of four people in Ireland live within a fifteen-minute walk of a shop, and two-thirds within fifteen minutes of a pub, restaurant or bus stop.[7]

The progress and fate of some recent and planned road transport projects is worth examining. The upgrade to the Dunkettle interchange, at the edge of Cork city, was finally completed in February 2024, at an eye-watering cost of €215 million. Taoiseach (and Corkman) Micheál Martin praised it as 'an extraordinary engineering achievement'. The upgraded junction would, Martin gushed, be 'a catalyst for economic and social activity' across the entire region.

The Dunkettle honeymoon was brief indeed. Within just five months, in the face of complaints from the public and from East Cork councillors who were repeatedly caught in gridlock at the interchange, Transport Infrastructure Ireland admitted that the upgrade was 'no silver bullet', adding that congestion could be solved only by . . . better public transport.[8] This followed multiple complaints from motorists about significant delays in accessing the Jack Lynch Tunnel.

The poor return on the huge outlay at the Dunkettle interchange in no way dampened political enthusiasm for big road projects. In June 2024 Transport Infrastructure Ireland (TII) announced a full upgrade to motorway of the

Cork–Limerick N20 at a cost of €2 billion.[9] A better use of a multi-billion euro investment would be to provide a direct rail link between Cork and Limerick, the second and third largest cities in the state. Today, this route requires passengers to change at Limerick Junction, wait for a connecting train and continue the dog-leg journey that takes an hour and three-quarters – ridiculously slow for travel between two cities that are just a hundred kilometres apart. An electrified high-frequency direct train link would easily cut the journey time in half. Remarkably, the 'All-Island Strategic Rail Review', published in July 2024, which offers proposals that it claims will transform our creaking rail network, entirely omits the option of a direct Cork–Limerick rail link.[10]

Road projects are a perennial political favourite. As junior minister Michael Healy-Rae said in early 2025, 'the people want tar and that is what we will give them'.[11] On the other hand, even modest attempts to facilitate public and active transport encounter intense (and often perverse) resistance from vested interests. The Dublin City Centre Transport Plan, developed by Dublin City Council, underwent an extensive period of public consultation in late 2023, and received over 3,500 written submissions from the public and interest groups, with more than 80 per cent in support of it.[12] The plan is intended to dramatically reduce the number of cars that enter the city centre purely in order to cut across it, and to double the share of journeys made by bicycle. The overall aim of the plan is to wrest back some of the public realm from traffic to make the city a safer, quieter, more pedestrian-friendly place, with landscaping, protected cycle lanes, widened footpaths and more pedestrianized zones, in line with many comparable European cities.

The Dublin Chamber of Commerce, representing around a

thousand businesses, reckons that traffic congestion is costing the city €350 million a year. Who could possibly want to stymie efforts to fix this absolute mess? The first sign of serious push-back came in May 2024, months after the consultation period had closed, when Ibec, the lobby group representing large businesses (which had not sent in a submission to the public consultation), released a statement calling for an 'immediate pause . . . to allow for an urgent examination' of the plan.[13] Ibec's last-gasp intervention was described by councillors as 'making a mockery' of the consultation process.[14]

Fine Gael junior minister Emer Higgins then intervened, publicly backing calls to delay the transport plan, even though transport was not within her ministerial remit. Next, an economic assessment commissioned by a group called the Dublin City Centre Traders Alliance – which includes five car-park owners – claimed improbably that the plan would reduce retail spending in the city by €141 million and cost nearly 1,800 jobs.[15]

In July 2024, Donald McDonald, CEO of retail giant Brown Thomas Arnotts, warned of the 'irreparable damage to city centre trade' and thousands of job losses that would arise from the transport plan.[16] The sky did not, after all, fall in following the imposition of traffic restrictions. Just five months later, after the restrictions were introduced, McDonald was enthusing about the 'electric' atmosphere in Dublin city centre over Christmas, with sales up 5 per cent on the previous year. 'Right across the board, every one of our stores were up on last year, and particularly Dublin city centre, which traded very well,' McDonald said.[17]

*

I went to college in UCC. There was no direct bus or train link in the 1980s from Kilkenny to Cork, so each journey involved first getting to Waterford city, either by hitching or by train, then catching the Cork bus.

Returning to college after Easter one year, I arrived late in Waterford and missed the only bus that day. With darkness approaching, I had to try hitching to Cork, but only made it as far as an isolated crossroads some distance from Youghal that night. There was a pub at the crossroads and I explained my dilemma to the publican, who was sympathetic but suspicious, and understandably wouldn't allow me to sleep in the pub lounge. He directed me instead to a barn at the back of the pub, where I clambered onto some bales of hay and eventually fell off to sleep. I hitched the last leg of the journey the following morning, probably looking somewhat bedraggled.

Like so many others, the first thing I did as soon as I started working was to buy my own wheels, starting with a decrepit Honda 50 and graduating through a variety of motorbikes until I was finally able to afford a car. The public transport alternatives at the time were both expensive and, frankly, dismal.

Freedom doesn't come cheap. The annual cost of running a car in Ireland, according to the AA, is over €10,500.[18] For many working people, this is a huge financial burden. According to the latest Central Statistics Office (CSO) household budget survey, the average Irish household spends almost as much on transport as on housing and food. Forced car dependence is widespread in rural Ireland and, with some 440,000 one-off houses scattered like confetti along our highways and byways, this is the daily reality for well over a million people for whom there is no ready alternative.

Car dependency has taken a dramatic toll on children. In 1986, just one child in four was driven to school. By 2022, the figure was six out of ten.[19] Apart from the negative impacts on children's health and fitness and on their ability to develop personal independence and resilience, early car dependency also makes them less likely as adults even to consider using public transport or to undertake active travel.[20] Today, one in five primary schoolchildren in Ireland is obese or overweight, compared with one in eight in 1990.[21] Each additional hour a person spends in a car each day is associated with a 6 per cent increase in the likelihood of obesity.[22]

A major report by the OECD argued that Ireland's Climate Action Plan, with its strong focus on electric vehicles rather than a fundamental shift towards active and public transport, might actually reinforce car dependency and all its associated problems. The report observed that alternative futures for car ownership and occupancy 'are simply absent from standard national transport modelling'.[23] Two-thirds of cuts to transport emissions are expected to be achieved as a result of the electrification of vehicles and the use of cleaner fuels. The government strategy, in other words, has been designed around the somewhat fatalistic assumption that the dominance of private vehicle use is both inevitable and likely to continue unchecked this decade and beyond.

After a very slow start, the uptake of EVs has risen sharply this decade. In 2019, there were only around 3,500 fully electric cars (termed 'battery electric vehicles', or BEVs) on Irish roads. Helped by a €5,000 grant to buyers (since reduced to €3,500), by 2023 BEV sales represented 19 per cent of the market, while hybrids and plug-in hybrids accounted for another 25 per cent.

In 2024, however, the rate of BEV uptake tailed off, accounting for just 14 per cent of sales. The slowdown can partly be explained by the fact that the rush of early adopters to this new technology is subsiding, and it's a harder slog to persuade less-motivated people to consider buying an EV. There has also been a slew of negative publicity around EVs, much of it inaccurate.

A major drag on EV uptake is the poor quality of the public charging network, much of which was installed by the ESB a decade and a half ago. The ESB currently operates around 1,600 public charge points nationally, but many of these are fitted with obsolete 22kW chargers that are too slow to be of practical use.[24] There are also some charging stations operated privately, by companies including Tesla and IONITY, but the quality and capacity of the network are nowhere near what they need to be.

I took the plunge and bought an EV seven years ago, a second-hand Tesla. (My purchasing decision was made long before Tesla's owner emerged as a far-right extremist.) At the time, it was the only car on the Irish market with a reliable range of around 350 kilometres, and Tesla offers the advantage of operating its own high-speed charging network that is simple to use, both in Ireland and abroad. I traded it in for another second-hand EV three years later, this one with a range of around 460 kilometres.

Even with our long-range EV, a recent trip to the Dingle peninsula was a hair-raising experience, with the handful of ESB charging points either low-speed or, in some cases, out of order, and we ended up almost completely dependent on a single EV charger at our hotel, leading to some heart-in-mouth moments. In July 2024, transport minister Eamon Ryan announced seventeen new high-power charging pools across Ireland, providing an additional 131 recharging points along major roads. Given the tens of thousands of new EVs being added to the national fleet every year, the scope of this project seems to be short by an order of magnitude. We need to be installing thousands of high-speed public charging points by the end of this decade, not hundreds.

Every innovation provokes a backlash, and so it has been with EVs. Much of this is manufactured rage, largely emanating from think tanks allied to fossil-fuel interests.[25] A host of myths and misconceptions about EVs now abound. One suggests that since more emissions are involved in manufacturing an EV versus a regular car, it is therefore not a 'greener' option. The truth is that the full life-cycle emissions from an EV are typically 52–69 per cent lower than from an equivalent petrol-engined car.[26]

Other myths include that EVs are a fire hazard or that the battery will have to be fully replaced after only 100,000 kilometres. In reality, EVs are twenty times *less* likely to catch fire than their internal-combustion-engine (ICE) counterparts.[27] Almost all EV manufacturers offer a full eight-year or 160, 000-kilometre warranty on the battery; this is significantly longer than the equivalent warranty on most ICE engines. Real-world experience suggests EV batteries will still be performing efficiently up to 300,000 kilometres, which is nearly twenty years of typical Irish driving.[28] An Irish start-up company called Range Therapy specializes in upgrading old Nissan Leaf EVs, replacing the 24kWh battery with a much more powerful 70kWh unit, which trebles the car's range.[29] Rather than being scrapped, the old 24kWh battery is then repurposed to be used as a storage unit for electricity from domestic solar panels.

The EU is committed to banning all sales of new ICEs by 2035, despite strong pushback from the European motor industry, which has been slow to transition to EVs and has fallen well behind China in terms of technology. The 2020

Programme for Government committed Ireland to banning the sale and import of new or second-hand diesel cars by 2030, but this commitment quietly disappeared from the 2025 Programme for Government.

Globally, battery prices continue to fall. While already cheaper to run – apart from much lower fuel costs, EVs have far fewer moving parts than ICE cars and don't require routine servicing – EVs are fast approaching price parity with fossil-fuel powered cars.[30] Already, the ten-year total cost of ownership of an EV in Ireland is between €12,000 and €18,000 less than its ICE equivalent.[31]

Another major benefit of EVs is that they are in essence batteries on wheels, and can actually help ease the pressure on our national grid by storing excess renewable energy when it is cheap and plentiful, then releasing it back to the grid in times of peak demand. All new public charging points will have to be enabled for smart charging, which will allow EVs to export energy back to the grid. A recent CCAC report urges that all residential charging points should also be capable of this.[32] Many already are.

If Ireland had, say, 500,000 EVs connected with bidirectional charging, and each of those were available to supply electricity back to the grid, that would be the equivalent of having two or three gas-fired power stations on standby. Peak-demand spikes may sometimes last only minutes, but at present the only way to avoid the risk of blackouts is to have excess (usually gas-fired) power on the grid 'just in case'. The total amount of battery backup available could be given a major boost if the national EV fleet were enabled to behave as a 'virtual battery', with EV users paid a premium rate for whatever energy they agree to sell back to the grid.[33]

*

While EVs represent a dramatic advance on their oil-powered antecedents, one thing they share with their ICE counterparts is the problem of bloat. Put simply, cars are getting bigger and heavier.

All the gains in improved fuel efficiency, as well as the fuel savings from EVs and plug-in hybrids, have been more than wiped out by the supersizing of cars globally. A European study found that cars are growing wider at the rate of one centimetre every two years.[34] Roads are not getting any wider, and with growing pressure to share road space with active travel, the proliferation of hulking great SUVs on our narrow streets and roads is creating increased hazards for both pedestrians and cyclists.

Despite the overwhelming evidence of dangerous levels of emissions, air pollution and noise associated with our car-choked streets and cities, efforts to reallocate road space towards public and active transport often encounter fierce resistance. A large-scale UK study identified what it described as people's unconscious biases due to cultural assumptions about the role of private cars, a phenomenon the researchers termed 'motornormativity', also known as 'car brain'.[35]

There is ample evidence not just that driving alters human behaviour, but that the size and value of the car play a decisive role. A US study found a strong correlation between the cost of a car and whether the driver yielded to let a pedestrian cross the street. For every $1,000 increase in the price of a car, the driver was 3 per cent less likely to yield to pedestrians.[36] Separate studies have found that driving larger cars makes the driver feel safer and more important, and this sense of personal safety encourages them to take more risks on the road.[37]

*

Since the public lost the battle for street space to cars a century or so ago, the fight has been ongoing to wrest some of it back. In September 1971, Dublin's Grafton Street was pedestrianized on a trial basis. Concerns that this would lead to 'hippies, street performers and vagrants in the street' led to the trial being abandoned.[38] It was not until December 1982 that the street was finally pedestrianized, a move that, decades later, has proven to be an unqualified success.[39]

Despite this, remarkably few other Dublin streets followed suit, with opposition from businesses and car-park owners among the chief reasons. Cork's Princes Street in 1971 became Ireland's first fully pedestrianized street. The city had to wait until the Covid lockdown of 2020 for a massive expansion, with seventeen streets permanently closed off to traffic in 2021.[40] Cork City Council says reaction to the move has been 'overwhelmingly positive', and more pedestrianization is planned. Having spent three years in car-choked Cork city when in college during the 1980s, it was a revelation to return recently and enjoy strolling through a revitalized city centre area that has come alive with on-street activities.

While on a family holiday in France in 2019, we had the misfortune to have to drive through Paris. The traffic was relentless and it felt genuinely unsafe. I was hugely relieved to have made it across the city, including through some famously chaotic roundabouts, without having been involved in a fender bender of some sort.

Since then, things have changed utterly – and for the better. The prime mover behind the greening of Paris is Anne Hidalgo, the city's mayor since 2014. Her vision was to free this historic city from the iron grip of traffic congestion. The iconic Champs-Élysées, once known as the most beautiful avenue in the world, had been turned into an eight-lane eyesore. Under Hidalgo's guidance, the Champs-Élysées is being transformed into an 'extraordinary garden' at a cost of around €250 million, or roughly the same as the Dunkettle upgrade work in Cork city. Similar greening is in progress in the area around the Eiffel Tower.

The Paris changes are far from cosmetic. Some 1,450 kilometres of protected cycle lanes have been built since Hidalgo came to office, while many major roads in the city centre, including along the River Seine, have been fully pedestrianized. Speed limits in most of Paris have been lowered to 30kph. A ban on all diesel cars entering the city came into force in 2024, and by 2030 this will be extended to petrol cars. To help combat rising temperatures and urban air pollution, Paris aims to have 170,000 trees planted by 2026. The changes have met with sustained resistance from the car lobby, as well as protests by taxi drivers, but public support is strong and the city authorities have pressed ahead. In a test of the changing mood towards cars, Parisians voted to treble parking costs for SUVs, with safety and air quality the key considerations.[41] It now costs €18 an hour to park an SUV in the centre of Paris.

Paris is also on track to remove half of its 140,000 on-street car parking spaces, especially on narrow and residential streets. Every car space takes up ten square metres, so the public are being consulted on how they would like to use the 700,000 square metres being freed up by their removal. Options include trees and plants, children's playgrounds, public toilets and vegetable allotments. Thanks to sustained focus on active travel, nearly three times as many Parisians now use bicycles for trips in the city centre area than cars.[42] Social norms have shifted too, and all this – remarkably – has been achieved in only around five years.

Elsewhere, the small Spanish city of Pontevedra in Galicia, population 83,000, used to be plagued by through traffic. On becoming mayor in 1999, Miguel Anxo Fernández Lores banned most car traffic from the city, and pedestrianized and paved 30,000 square metres of its medieval centre.[43] Pontevedra's last recorded road-related fatality was in 2011, while air quality has improved by 67 per cent and 15,000 people have moved back into a city centre that had been in long-term decline.

A recurring theme of efforts to loosen the grip of cars on urban areas is how often such moves are initially unpopular.[44] In 2007, much of the centre of Ljubljana, the capital of Slovenia, was closed to cars, a move supported by only 40 per cent of the population. Just ten years later, a follow-up survey found 97 per cent of citizens 'opposed the reopening of the centre to motor vehicles'. Similarly in Graz, Austria, the introduction of a 30kph speed limit in the 1990s was bitterly opposed; once the measure was implemented, public support for it rose to over 80 per cent.[45]

This kind of transformation can be made in any urban area, but it is likely to encounter extreme resistance from people, including politicians, who are only able to see the world through the windscreens of their cars. An egregious example of populist pandering to the car lobby on active transport came from Fine Gael MEP Regina Doherty, who in June 2024 made the ridiculous claim that 'authoritarian' cycle lanes in Dublin 'have divided the city like East and West Berlin'.[46]

*

To begin the task of easing the chokehold of cars on our streets, we have to challenge their dominance in our media and in our culture.

The car industry spends an estimated €50 million a year on advertising in Ireland.[47] That money buys a lot of persuasion, and presents a genuine dilemma for Irish media outlets, for whom car advertising represents a significant slice of their revenues. Consider that RTÉ's flagship early evening radio programme is called *Drivetime*: a name that is based on the assumption that we're all in cars between 5 and 7 p.m. Consider too that radio bulletins give regular 'traffic updates' (until recently supplied by the AA, an organization representing private car owners) focused entirely on delays affecting private cars. The bulletins rarely have anything to say about the progress and availability of trains, trams and buses.

It was a financial arrangement involving Renault that led to the end of Ryan Tubridy's employment at RTÉ in 2023. And the *Late Late Show*, under an eight-year sponsorship arrangement with Renault that was worth up to €750,000 a year, frequently staged competitions with cars as prizes.[48] In 2023 we also learned that a number of RTÉ stars were involved in lucrative 'deals for wheels' to act as 'ambassadors' for car brands.[49]

It doesn't have to be this way. France has implemented rules requiring car manufacturers to include one of three slogans in all TV, radio, print and online adverts: 'On a daily basis, take public transport'; 'Consider carpooling'; or 'For short journeys, walking or bicycling is preferable'. Ads must also include a hashtag that translates to 'Move and pollute less'.[50] France aims to ban adverts on cars producing more than 123 grams of CO_2 per kilometre by 2028, a measure that will hit larger petrol and diesel cars, especially SUVs, hard. The French government has already increased the top level of eco-tax on highly polluting cars from €30,000 to €50,000, sending a powerful signal to both motorists and the industry that the era of climate-wrecking gas-guzzlers is fast coming to a close.

Perhaps the most useful step a city like Dublin could take to address our car culture would be to follow the lead of Edinburgh. In 2024 the Scottish capital's city council announced a ban on ads for fossil-fuel products, including all SUVs and plug-in hybrids, cruise holidays, airlines and airports.[51]

Neuroscientist Clare Kelly is wary of efforts to place the onus of change on individuals altering their behaviour. Governments, she told me, 'love to focus on nudges because they don't restrict choice, they don't make things more expensive, and they're popular with voters'. However, the interventions proven to work involve system-level changes, such as restricting certain types of vehicles or taking from existing road space to build active-travel infrastructure.

Research published in 2024 by ESRI behavioural scientist Shane Timmons looked at the challenges of building the physical infrastructure and public support for the transition from cars to active travel.[52] It found that for people considering taking up cycling, personal safety concerns weighed most heavily. Perhaps the largest psychological barrier to active-travel infrastructure is the status quo bias, meaning the tendency of individuals to prefer things to remain the same, even if change may actually be beneficial.

The challenge of sharing finite road space between motorists, public transport vehicles and cyclists is also complicated by high levels of entitlement and aggression shown by some drivers towards vulnerable road users. A survey of 1,000 female cyclists in London found that 93 per cent reported drivers using their vehicles to intimidate them.[53] The aggression extends to pedestrians too. My daughter and a school friend were crossing the street in Dún Laoghaire some years back when a motorist deliberately accelerated and swerved towards them, forcing them both to sprint onto the footpath. He had made eye contact with them throughout and was presumably doing it for his twisted version of fun.

Open hostility to cyclists is not restricted to apoplectic drivers. It permeates media coverage too, some of which positions drivers as victims. In an *Irish Times* review of a new Toyota SUV, Michael Sheridan wrote of the benefits of the car's high seating position: 'Drivers can now interact with more vulnerable road users like cyclists and pedestrians at near eye level. So you can look your abusers in the eye.'[54]

I shared a platform at an HSE conference on climate and health in Dublin in late 2024 with Seán Owens, a Dundalk-based GP who is involved in Irish Doctors for the Environment. His contribution was electrifying. I was due to join him at another climate event in January 2025, but he couldn't attend. Owens had recently written an open letter to local county councillors with the subject line 'Why Are We in Such a Hurry to Kill Ourselves . . . and the Planet?'[55] His plea was for lower speed limits to make the roads safer for cyclists and pedestrians. On his way home from surgery early one evening that January, Dr Owens was knocked off his bike by a hit-and-run driver and left critically injured. He faces a long and extremely difficult recovery.

Challenging the dangerous culture of aggressive entitlement among some car users is a crucial first step to safely sharing our limited road spaces.

*

Moving the dial on transport emissions while also tackling congestion and freeing up valuable urban space requires a decisive shift towards public transport. A train from Dublin to Cork takes ninety cars off the road and saves on average 1.6 tonnes of CO_2 per journey, even when the train is diesel-powered.[56] Electrifying all our rail networks would further cut transport emissions.

An important milestone for public transport in Ireland occurred in January 2025 with the introduction of new twenty-four-hour bus routes in Dublin.[57] I've lived, worked and socialized in and around Dublin for nearly forty years, and in all that time it has been a source of bewilderment that the buses stopped running by midnight and the DART service, ridiculously, shuts down at around 11.30 p.m. every night, leaving thousands of people dependent on expensive and often unreliable taxis. Twenty-four-hour public transport is essential for any European capital city.

Rural Ireland has long been the toughest public transport nut to crack, but there has been real progress in recent years. Transport For Ireland (TFI) has been rapidly building and expanding its Local Link bus service for towns and townlands in rural areas. In 2021, it was carrying around 16,700 passengers a week; by 2024, this had risen six-fold, to over 100,000 weekly passengers. More than 190 towns and villages are now connected by Local Link services.

I have mentioned the failure of 2024's 'All-Island Strategic Rail Review' to address the fact that there is no direct rail link between Cork and Limerick – the second and third largest cities in the state, separated by a mere hundred kilometres. Although the measures that the €37-billion plan does propose – including increasing the speed and frequency of intercity services and reopening numerous shuttered lines – do not go nearly far enough, they will help make the rail network more efficient and more attractive to users. In total, an additional 650 kilometres of line are planned, to connect towns currently without rail links. The plan also envisages new rail links for Dublin Port and for Shannon and Dublin airports. This move would bring an additional 700,000 people to within five kilometres of a station with a regular train service.[58]

In the greater Dublin area, the long-delayed underground MetroLink project involving sixteen stations, linking with the existing DART network and running from the city to beyond Swords, will be complemented by the DART+ programme.[59] This will see the DART network expand from its current fifty kilometres to over 150 kilometres, and promises to promote multimodal and active transport. The annual investment required would be equivalent to what Ireland was spending on developing its motorway network in the late 1990s and early 2000s. While the 2025 Programme for Government 'fully commits' to implementing existing proposals for rail upgrades, it remains to be seen whether a government propped up by tar-loving rural independents and without a Green Party involvement will actually honour its commitments on public transport.

Train travel is unloved in Ireland, and no wonder. The current level of service on the rail network is abysmal, with its lack of connectivity, poor-quality rolling stock, no catering on most services and timetables that make no sense whatever. Want to get the train from Killarney to Limerick, or Ballina to Sligo? Forget about it. The biggest planned new route runs from Mullingar to Portadown, via Cavan, Clones, Monaghan and Armagh. While welcome, this hardly seems like a route crying out for a rail link.

To be effective, public transport depends on the network effect: the better it gets, the more people use it, leading to a virtuous cycle of investment and usage. Perhaps the biggest battle ahead will be in restoring public confidence in rail as a reliable, well-connected travel alternative. This will happen only if the state enacts the improvements planned in the 'All-Island Strategic Rail Review' – and then goes much further.

*

One-fifth of all emissions in Ireland's transport sector are from heavy goods vehicles (HGVs).[60] Ireland's relatively short distances mean our HGV fleet is well suited to being upgraded to battery-electric. The Department of Transport opened a purchase grant scheme in early 2024 to assist haulage firms to switch away from diesel and to invest in battery-powered HGVs. The lifetime running costs of electric HGVs are dramatically lower than diesel, which should make them attractive to road haulage firms, especially when teamed with commercial solar PV installations.

The vast bulk of freight in Ireland is moved by road, with impacts in terms of emissions, air pollution and congestion. Apart from cleaning up the existing fleet of HGVs and light commercial vehicles, there is also an imperative for a shift back to using our rail system to carry a far greater share of freight than at present. Rail is inherently much more fuel-efficient at moving goods than trucks.[61]

In 1981, around 4 million tonnes of freight were carried on Irish railways. Four decades later, that had plummeted to around 300,000 tonnes, under 1 per cent of the freight transported in Ireland in 2022. This is by far the lowest share of freight carried by rail of any European country.[62] As an island, we are disconnected from European rail networks; but Norway and Sweden, while also geographically remote from most of the European mainland, carry 14.5 per cent and 28.7 per cent respectively of all their freight by rail.

Irish Rail's rail freight strategy sets a target of around

10 per cent of freight travelling by rail. Key to achieving this will be establishing rail freight connections to the ports at Dublin, Foynes and Cork.

*

If you're looking for signs of meaningful action to tackle the climate emergency, one place you assuredly will not find them is in aviation. If the aviation sector were a country, it would be among the world's top ten carbon polluters. Bizarrely, emissions from international aviation are excluded from Ireland's national emissions reporting.

In just the last three decades, Irish aviation emissions have grown by 500 per cent.[63] One key reason for the explosive increase in aviation, both nationally and globally, is that the sector is massively subsidized and pays almost nothing towards the pollution and climate damage it inflicts.

Jet kerosene for commercial aviation is exempt from both excise and carbon taxes in the EU.[64] On top of these lavish subsidies, the Irish state pours significant resources into aviation, including €35.6 million in regional airport funding in Budget 2023.[65] What's more, airline tickets are VAT-exempt in the EU, as is aircraft leasing – providing further hidden subsidies to the sector.[66]

These subsidies create market distortions and perverse incentives. For instance, flying from Kerry to Dublin can be cheaper than taking the train, or even the bus. The vastly more environmentally sustainable train has to pay an effective carbon tax of €50 per tonne of fuel, a cost the airlines don't have to bear.

While aircraft have become significantly more fuel efficient in the last three decades, any emissions savings have been more than wiped out by the rapid growth of aviation. While other parts of our transport system have realistic, if challenging, pathways towards major emissions reductions, aviation is operating on a wing and a prayer; apart from a vague faith in expensive and elusive future technologies such as sustainable aviation fuels (SAFs), the industry simply has no plan to reduce its climate impact. At present, total SAF production can meet less than 1 per cent of the global demand for aviation fuel, and the SAF that is available is nearly three times more expensive than its fossil-fuel equivalent.[67]

Apart from its oversized impact on the global climate, what is most egregious about aviation is its sheer inequality. Any of us who have ever set foot on an aeroplane are highly privileged, given that more than four in five people in the world have never flown. Within the small sector of humanity that uses air travel, there is a subset that is disproportionately responsible for the damage done by aviation: just 1 per cent of the population in wealthy countries takes around 50 per cent of all flights.[68]

The ultimate super-emitters are those who use private jets. In 2022, nearly 6,700 private jet flights departed from Irish airports. To underline the sheer vanity of private aviation, bear in mind that the most common destination from Ireland was London, despite the fact that there are around forty commercial flights available from Dublin to London every day.[69] People using private jets are responsible for up to fourteen times more carbon pollution per person per journey than those on commercial aircraft.

The Irish state owns DAA, the semi-state company that operates Dublin and Cork airports. Although it is state policy to reduce GHG emissions across all sectors, DAA has been involved in a long-running campaign to undermine this objective. It is lobbying hard to have the annual cap of 32 million passengers through Dublin airport lifted to allow up to 40 million passengers a year, which would, according to DAA's own planning application, translate into a 22 per cent increase in carbon emissions.[70] DAA's chief executive, Kenny Jacobs, a former Ryanair executive, claimed in 2024 that the emissions increase would be 'minimal . . . if any', owing to the use of more efficient fuel and aircraft. When pressed to explain his comments, Jacobs added: 'I'm not rejecting science. I'm supporting the economy.'

Jacobs was backed up by the junior transport minister at the time (and later finance minister), Jack Chambers. 'Lifting a cap and facilitating more flights is in the context of more fuel-efficient aircraft, really ambitious targets on sustainable aviation fuel, which we have to meet in an EU and international context,' he said.[71] It takes quite a leap of logic to square more flights with fewer emissions.

To bring aviation in line with the urgent need to cut carbon emissions, the first and obvious step would be to tax it globally as other transport fuels are taxed, as well as applying hefty carbon taxes to reflect the true costs of flying. In addition to taxing aviation fuel in line with its climate impacts, I believe a fair and equitable solution to the gross overuse of aviation by a small minority would be to introduce a system of flight rationing. I suggest that each individual be allocated a certain distance, say 1,750 kilometres, annually (equivalent to a return flight from Dublin to Paris, producing around 210 kilograms of CO_2), with this non-transferrable allocation tied to your PPS or passport number. If you don't take any flights in a given year, your allowance can be carried forward to the next year, but it cannot be sold or traded. Scaled up nationally and applied equally to all 5.2 million Irish citizens, that means around 1 million tonnes of aviation emissions would be exempt from additional climate levies. For the rest — and Ireland's international aviation emissions totalled 3.43 million tonnes in 2023 — the age of dirt-cheap flights would be over.[72]

If you took another similar return flight, your next 1,750 kilometres would attract, say, a €250 climate levy. From there, the levy doubles with every additional round trip: €500, €1,000, €2,000 and so on. Eventually, the cost will start to deter even the wealthy. This would obviously hit those who fly regularly for business, usefully challenging companies to think much harder about how many of these flights are really needed.

A system like this would require some flexibility around, for instance, compassionate grounds in the event of a bereavement, but these rules would have to be rigorously enforced to be effective, with a high burden of proof placed on the person applying for an exemption. Properly managed, it could be hugely effective in clipping the wings of frequent flyers. Lasting change will come only when the price of an airline ticket reflects the true cost of flying.

5. The New Land War

In July 2023, three EPA scientists were invited to give expert testimony to the Joint Oireachtas Committee on Agriculture, Food and the Marine regarding the agency's latest water quality monitoring report. It pointed to the fact that the nitrates derogation, the 'temporary' EU exemption allowing many Irish farms to carry nitrogen at well above the scientifically established safe levels, was contributing to a deterioration in water quality.

The committee chair, Fianna Fáil TD and dairy farmer Jackie Cahill, went on the attack against the scientists from the outset. He was joined by an independent senator, Victor Boyhan, who claimed the EPA report was 'not scientific and its integrity is questionable' – an outrageous slur on the agency and its scientists.

Also on the attack was independent TD Michael Fitzmaurice, who suggested that a river subject to ongoing pollution can, as if by magic, 'clear itself out'. Sinn Féin TD Martin Browne and Fine Gael's Michael Ring also attempted to undermine the evidence being presented, with Ring describing the EPA itself as a 'necessary evil'.

After the EPA delegation had left the committee room, Cahill, as committee chair, summed up as follows: 'it was clear that in spite of significant increases in dairy cow numbers since 2015, water quality has not disimproved'.[1]

This was a complete misrepresentation of everything the EPA scientists had just said, but there was nobody in the

committee room to call Cahill out on this, as the EPA delegation had already left. While Ireland's industrial livestock
lobby and its political allies have long deployed incendiary
language against environmental NGOs, it was disturbing to
see elected officials make political attacks on the state's own
regulatory agency.

That Oireachtas committee hearing was far from an isolated incident. For instance, in February 2023, a draft EPA
Land Use Review document was described by the IFA as
'fundamentally flawed', with its then president, Tim Cullinan,
threatening an 'uprising in rural Ireland'. The draft report
had indicated that Ireland would need to reduce its total
livestock numbers by 30 per cent, as well as greatly increasing forestry cover and rewetting up to 302,000 hectares of
peatlands drained for farming.[2] Fitzmaurice labelled this as
amounting to an attempted 'ethnic cleansing' of the agricultural community, adding that the report would do more
damage to rural Ireland than Cromwell had. 'At least Cromwell didn't mind us living in Connacht. These don't seem to
want us to live anywhere.'[3]

I faced off against Fitzmaurice on a TV panel discussion
some weeks later, and asked him if he would be prepared
to withdraw his scurrilous analogy with 'ethnic cleansing',
pointing out that the phrase means the mass expulsion or
killing of an ethnic group. It is not, in other words, a phrase
to be tossed around lightly. His response: 'I will never apologise to you or your like.'[4]

*

Across Ireland, the dappled hay meadows of my youth have
all but disappeared, replaced by the bright monochromatic
green of ryegrass for pasture and silage. Since the mid-1950s,
hay production has declined by around 90 per cent.[5] While

this may have been good for productivity, it was very bad news for the biodiversity, including farmland birds, that once teemed in hay meadows.

In the course of the last half-century, European farmland bird populations have been in sharp decline, with 57 per cent disappearing largely as a result of a combination of mechanization, pesticides and land clearance for intensive crop production. In Ireland, the situation is significantly worse, with 63 per cent of our bird species in decline and one in four now seriously threatened.[6]

Looks can be deceiving. To the casual observer driving or travelling by train through rural Ireland, our countryside appears as famously green and lush as the cinematic landscape that the actress Saoirse Ronan strode through in a lavish 'Origin Green' advert developed by the state food marketing agency, Bord Bia, in 2012.[7]

'We did not inherit this world from our parents; we borrowed it from our children,' was the emotive opening line delivered by Ronan. Ireland, she whispered, 'can become a world leader in sustainably produced food and drink'.

Why? Because, apparently, 'our climate has always been this mild, our landscape this lush, our fields have always been this green, and windswept, and rain-washed . . . the elements have always conspired here to make great farming possible'.

Like all good yarns, it contains a kernel of truth. Ireland has indeed been blessed with a temperate climate and plenty of rainfall. But to suggest that today's monocultural landscapes of intensively managed ryegrass and precious little else harkens back to some primordial, unspoiled Ireland is pure marketing dross. And it is especially nonsensical to invoke our eternal and unchanging climate in an era of rapid climate change.

This beautifully produced advert was a textbook example

of greenwashing, not dissimilar to the fossil-fuel industry's absurd efforts at creating brands such as 'clean coal'. The dairy intensification being promoted was ironically in the process of destroying any remaining vestiges of the idyll the advert was evoking.

Bord Bia claims that 61,000 farmers and over 300 'leading Irish food and drink companies' are signed up to Origin Green.[8] In 2023, Origin Green listed ninety-seven Gold Members – companies that, it says, have shown exceptional environmental and sustainability performance. In fact, numerous Gold Members, including Kepak, Dawn Meats and Dairygold, have been successfully prosecuted by the EPA for environmental offences in recent years.

You might imagine that environmental sustainability, which is at the heart of Bord Bia's mission statement, would be reflected in the composition of its board. This is manifestly not the case. Of the twelve board members listed, there are a total of zero ecologists or similarly qualified people.[9] From 2018 to 2024 its chair was Dan MacSweeney the former chief executive of Carbery, a dairy and nutrition company. In September 2017, while MacSweeney was still in charge, Carbery was placed on an EPA watchlist for environmental breaches.[10] A current board member is the chief executive of Dawn Meats, which the EPA successfully prosecuted in 2023 for a series of environmental breaches, including providing the EPA with 'information which was false or misleading' regarding environmental monitoring.[11]

Livestock is in the very DNA of Bord Bia, which has its origins in the Irish Meat and Livestock Board.[12] Its continuing orientation towards 'meat and livestock' became embarrassingly clear when, in 2018, Bord Bia issued a tender looking for consultancy firms to help it 'win back' vegetarians and vegans.[13] The process was highly revealing about the structural bias within state agencies in favour of high-emission and highly polluting forms of agriculture over lower-impact plant-based food systems.

A few years ago, I asked Bord Bia what percentage of farmers who applied for Origin Green certification under its 'Sustainable Dairy Assurance Scheme' (SDAS) were successful. The answer was around 99.5 per cent. Either we have the world's most uniquely green dairy farming community, or Bord Bia's scheme is extraordinarily lax in its criteria. I revisited this issue with Bord Bia recently. I was told that 62 per cent of dairy farmers enrolled in its SDAS were found to be in 'major' non-compliance in 2024. Despite this massive failure rate, there is 'no de-certification arising from a non-compliance being raised during subsequent renewal audits, provided such issues are "closed out" after the audit', Bord Bia confirmed to me.

Over the past decade, Bord Bia's taxpayer-funded budget has more than doubled. It is instructive to see how this ever-expanded budget is being deployed. In September 2022, Bord Bia rolled out an €8-million campaign, co-funded by the EU, that was aimed at growing Irish food exports to Asia. The programme being spearheaded featured three of the world's most emissions-intensive foods: beef, dairy and lamb.[14]

Agriculture minister Charlie McConalogue stood proudly in front of a banner in Tokyo that read 'Sustainable beef and lamb from Ireland'. Bord Bia highlighted one outlet going the extra mile – and more – for sustainability: Singapore-based Ryan's Grocery. It imports chilled Irish Angus beef weekly by aircraft and markets its premium 'sustainable' Irish striploin steak at a cool €72 per kilo.[15]

Just days before the high-powered Irish delegation took flight for the Asian sales jamboree, a ship containing 33,000 tonnes of Ukrainian grain was unloaded at the port of Foynes in Co. Limerick, having beaten the Russian blockade. The entire shipment was for importer R&H Hall, to be sold as animal feed.[16]

By selling high-emissions luxury food to the rich while fattening animals with grain that could otherwise feed hungry people in poor countries, Ireland is now a major international exporter of food insecurity. But to listen to our government ministers and semi-state bodies, you would never know it.

*

The potency of methane as a greenhouse gas is extremely inconvenient to the livestock industry. As a result, it has launched a sustained counter-offensive, promoting misinformation aimed at deflecting from the urgent need to curb methane emissions by reducing ruminant herd sizes.

The industry's most plausible line of argument is that technological fixes, such as dietary additives, selective breeding or better grassland management, can mitigate emissions. In Ireland, Teagasc has in recent years carried out extensive research looking at methane mitigation strategies, but none has yet delivered meaningful cuts in emissions for pasture-based farming. The one solution we know for certain delivers results – reducing herd sizes – is no longer even alluded to in Teagasc's extensive published work on methane reduction.[17]

In July 2022, Frank Mitloehner of the University of California, Davis, testified before the joint Oireachtas committee on agriculture. This was not his first appearance in Dublin: he'd been a guest speaker at an IFA-hosted event in 2020. Mitloehner told the committee that, in the state of California, through the use of anaerobic digesters, 'we have reduced 30 per cent of the dairy sector's methane over the past five years'.[18] Mitloehner went further, claiming that these 'drastic' and 'amazing' results meant there were technological solutions to the livestock methane problem that 'many believe can only be tackled through draconian herd size reductions and dietary changes'.

Committee chair Jackie Cahill told RTÉ radio the following January that there was no need to reduce Ireland's national herd, based on Mitloehner's evidence to the Oireachtas committee. 'We see in California how they reduced their emissions by 30 per cent, so the technology does work,' he stated. But in the meantime it had been pointed out, with reference to California state data, that Mitloehner's claim was simply false. When interviewed for the same report, Mitloehner walked back the statements he had made repeatedly to the committee the previous summer:

> The 30 per cent is not a reduction of total emissions . . . it's 30 per cent of the reduction goal; so the [California] dairy industry needs to reduce 7 million tonnes, they have so far reduced 2.2 million tonnes, which is 30 per cent of the reduction goal . . . when I saw that this was misinterpreted, I added a written comment to my statement and I think I was very clear about it.[19]

Given his own stated awareness of the need for precision, Mitloehner failed to explain why he repeatedly gave ambiguous, incomplete and misleading information to the Oireachtas committee, stressing 'amazing' results that were, in reality, far from amazing. Nobody who listened to his Oireachtas committee contributions was in any doubt about what he had said on the public record.

In the same RTÉ interview, Mitloehner also confirmed that 'if you increase methane over time, then you are increasing warming, and quite dramatically so'. Teagasc's dairy roadmap for 2027 sees milk output increase by 1.5 billion litres, which, as Mitloehner admitted, means 'quite dramatic' additional global warming.

Mitloehner's pitch to the Oireachtas committee has made it much easier for the Irish livestock sector to defend its business-as-usual pathway. The EPA had indicated that methane levels in agriculture need to fall by around 30 per cent to keep the sector on track for even the lower end of its emissions targets; Mitloehner's supposed 30 per cent methane cuts in California, driven by technological tweaks rather than herd reduction, would seem to square the circle.

Who was this Frank Mitloehner, crossing the globe to show Ireland the way in agricultural methane reduction? A major Greenpeace investigation in late 2022, which was cited on the front page of the *New York Times*, drew attention to Mitloehner's livestock industry links and funding. The centre he runs at UC Davis had its structure agreed via a memorandum of understanding between the university and a subsidiary of the American Feed Industry Association.

In exchange for millions of dollars in livestock industry funding, UC Davis's CLEAR Center maintains what the Greenpeace investigation described as an advisory board comprising '12 of its agribusiness funders, to provide "input and advice" on the "research and communications priorities of the industry"'.[20]

Despite the narrative to the contrary, Ireland's contribution to the climate emergency is anything but trivial, and livestock expansion is at the very heart of this crisis. In terms of avoiding the very worst impacts of climate breakdown, what happens now and in the next decade is absolutely crucial.

Efforts to wish away the serious global-warming impacts of methane made it into the 2025 Programme for Government,

which recognized what it senselessly called the 'distinct char-
acteristics of biogenic methane' while also promising to
advocate for methane to be reclassified at EU and global
level. This was just one of a slew of concessions to the livestock sector by the rural-tinted incoming government.[21]

*

Denial of climate science is a global industry, but Ireland was
at least temporarily its epicentre in October 2022, with the
publication of the so-called 'Dublin Declaration'.[22]

Then junior agriculture minister (and beef farmer) Martin
Heydon formally launched the Declaration at the International Summit on the Societal Role of Meat, which was
hosted by Teagasc at the Food Research Centre in Ashtown,
Dublin, with European commissioner Mairead McGuinness
along to add support.[23] The event cost €45,000, of which
€39,000 was provided by Teagasc.[24]

It received widespread coverage in the Irish media, including news reports on RTÉ and in the *Irish Times*. Writing after
the event, Peer Ederer, a financial economist and one of the
authors of the declaration, noted, 'What started the coverage
in Dublin is not the brilliance of our science but the fact that
we had a minister and a commissioner there.'[25]

The aim of the Declaration was to defend the multi-billion-
euro global livestock industry, which it describes as 'too
precious to society to become the victim of simplification,
reductionism or zealotry'. If this wording sounds peculiar,
it's because it is the language of lobbying, not science.

A total of around 1,200 people are listed as signatories
to the Declaration, including twenty-seven from Ireland, of
whom sixteen were Teagasc employees.[26] Quite why employees of a state research agency would choose to put their
names to a manifestly political declaration like this is unclear.

An in-depth investigation a year after its launch revealed that the people behind the Declaration had deep ties to the livestock industry.[27] After Ederer was contacted by investigative journalists, a page was added to the Dublin Declaration website disclosing the various commercial interests of its committee members.[28] Declaration of actual or possible conflicts of interest is a keystone of scientific transparency and integrity. Here, it only occurred months after the event and in response to journalists' questions.

The Declaration was promoted by the Global Meat Alliance, an industry lobby that includes Irish meat corporations, with PR carried out by Red Flag, a Dublin-headquartered consultancy.[29] A 2018 Red Flag brochure boasted that its campaign and advice 'led to the World Health Organization (WHO) walking away from the International Agency for Research on Cancer's allegation that red meat was a serious cancer risk'.[30]

The Dublin Declaration is a case of the livestock sector taking a leaf out of the fossil-fuel industry's playbook in creating documents with the appearance of science. A follow-up meeting in Brussels in April 2023 to further promote the Dublin Declaration was opened by Frank O'Mara, director of Teagasc.[31] A 2025 ESRI report cited 'undisclosed conflicts of interest, some of which specifically concern Teagasc', in the Dublin Declaration.

An analysis of the Dublin Declaration was published in the journal *Environmental Science & Policy* in late 2024.[32] It concluded that the Declaration 'echoes meat industry narratives' and undermines scientific institutions that are communicating more accurately on the topic. It pointed out that Teagasc is, notably, 'invested in the Irish dairy sector and generates income from livestock trade and providing services to the industry', and urged that it seek to reduce its conflict of interests by 'divesting from the industry, or at least diversifying its portfolio towards non-animal based food producers'. Apart from its exchequer funding, Teagasc also receives around €20 million a year in advisory-service and course-fee income, and nearly €4 million in livestock-trading income from farms it operates.[33] The Irish public deserves an agriculture research agency that is fit for purpose, one that can be an honest broker between the needs of industry and wider society as well as state policy on key issues such as agricultural emissions reduction. As it is currently structured, Teagasc falls far short of that standard.

Teagasc is also an *ex officio* member of the CCAC and, as such, heavily influences the council's output relating to climate and agriculture. An independent review of the CCAC's operations warned that Teagasc was 'representing a sectoral interest . . . blurring the lines between independent expert and stakeholder'.[34] This clear advice that Teagasc be removed from the CCAC has been ignored, for what I have to assume are political reasons.

*

In May 2021, an article was posted on Fine Gael's website under the title 'An Taisce a Leading Threat to Future of Rural Ireland'. This post, which was still online at the time of writing, involved an all-out attack on Ireland's oldest charity, which has a specific remit in Irish law to review and comment on planning issues.

'The only interests An Taisce seems hell bent on safeguarding are those of a chosen few driving what seems to be a very personal vendetta against farmers, the rural economy and a company like Glanbia,' alleged the statement, which was attributed to six Fine Gael members of the Oireachtas:

four TDs (all of whom were former ministers) and two senators.

This outburst followed An Taisce's decision to appeal a High Court decision granting planning permission for a cheese plant in Co. Kilkenny that was a joint venture between Glanbia and a Dutch cheesemaker, Royal A-ware. An Taisce had argued that the 450 million litres of milk the new plant would consume would have negative consequences on both emissions and water quality. The Supreme Court rejected the appeal.[35] 'The government must now look again at An Taisce being funded by the taxpayer,' the Fine Gael statement continued.

Fianna Fáil's Jackie Cahill upped the ante, describing the appeal against the High Court's decision as a 'revolting act of treason'.[36] Both Cahill and another signatory of the Fine Gael letter, John Paul Phelan TD, held shares in Glanbia at the time of their incendiary outbursts.[37] (I was a volunteer member of the An Taisce Climate Committee at this time, and was on the receiving end of a string of personalized attacks and threats as a result of the toxic atmosphere whipped up around this issue.)

Joining in the pile-on, Glanbia chief executive Jim Bergin attacked state funding of environmental NGOs, which in 2021 totalled €1.8 million, or roughly a tenth of 1 per cent of the annual subsidies received by Irish agriculture.[38] In 2014 alone, Bergin received a bonus in the form of Glanbia shares worth more than half a million euros, which was more than twenty times the average farm income that year.[39] And in 2023, Glanbia's managing director Siobhán Talbot's total remuneration package came to €5.99 million, or more than three times the total annual state funding for the entire environmental NGO sector.[40]

In a moment of supreme irony, the following year, in April 2024, Bergin, chief executive of Tirlán (formerly Glanbia), asked the government to spend €40 million to help firms like Tirlán to 'solve the nitrates issue' by reducing pollution levels in the country's 'worst rivers'.[41] While Bergin was upset to see funding of less than €2 million for environmental NGOs working to protect our natural habitats, he had no problem demanding twenty times that amount to clean up the very mess that An Taisce and other NGOs had warned would occur as a result of the untrammelled intensive dairy expansion that the industry had lobbied tirelessly for.

Bergin went even further, arguing that the €3 billion set aside in the state's climate and nature fund established in Budget 2024 should be raided and spent on improving water quality.[42] So, billions in taxpayer funds dedicated to climate and nature should instead be funnelled into fixing the problems created in large part by the very industry Bergin represents. The end goal of this proposed massive taxpayer investment in water quality, according to Bergin, is so that the nitrates derogation, allowing farmers to use levels of nitrogen above the scientifically safe maximum levels, can be preserved. This is the Alice-in-Wonderland nature of agri-industrial lobbying in Ireland today.

The smaller farming organizations, perhaps lacking the IFA's easy access to media coverage, seem to work extra hard to gain attention. For instance, Dermot Kelleher, president of the Irish Cattle and Sheep Farmers' Association (ICSA), claimed in 2023 that 'a small cabal of unrepresentative but noisy activists were salivating at the prospect of ripping out the heart of economic activity in Ireland'.[43]

The future, Kelleher warned, was an 'insane vision of a tiny minority where wolves would roam a rural wasteland

and consumers would be forced to eat insect protein and fake burgers'. Insane vision indeed.

The secretary general of the ICSA, Eddie Punch, vented his frustrations to a meeting of farmers in Clare in 2021. 'I know you have to listen to the environmental side of it but I am sick of listening to hippy dippies and tree huggers telling us we need to do more for the environment.'[44] It was ugly but effective, making the lead story in the local paper and doubtless boosting Punch's political profile as well. (Punch was an unsuccessful candidate for the Independent Ireland party in the 2024 general election.)

Ahead of the 2020 general election, ICMSA president Pat McCormack claimed the biggest issue for dairy farmers was 'constant attacks on their livelihoods . . . by the most aggressive and arrogant elements of the environmental movement'.[45] Two years later, McCormack accused environmental groups of 'holding the national interest hostage'.[46] It is a recurring tendency of lobby groups to conflate their own narrow sectoral interests with those of the nation as a whole.

In a radio interview in 2023 reacting to the EU's proposed Nature Restoration Law, McCormack appeared to flatly reject the scientific evidence that there even was a biodiversity crisis in Ireland and across Europe, suggesting rather improbably that the real reason for any decline in biodiversity was somehow related to motorways or the impact of wild mink.[47]

There really is nothing at all normal about this sustained level of vitriol. This was made clear by senior European Commission official Aurel Ciobanu-Dordea, who warned about 'the increasingly aggressive stance being taken against environmental campaigners in Ireland'. While these kinds of attacks are widespread in authoritarian regimes like Hungary, Ciobanu-Dordea said it was 'highly unusual for an advanced society like Ireland to witness such conduct'. He added that 'negative reporting in mainstream media, even from politicians', is commonplace in Ireland, including threats to 'cut off funding to certain NGOs'.[48] Given how we have seen democratic norms crumble across the world, Ireland should not be complacent in tolerating repeated politicized attacks on the handful of organizations, mostly run by volunteers and operating on a shoestring, which are fighting on behalf of us all to protect nature, reduce pollution and rein in climate change.

*

The relentless mud-slinging by some in the Irish agri-industrial lobby has not completely succeeded in swaying the general public. Research carried out by the National Dairy Council (NDC) in 2022 found that two in five consumers disagreed with the statement 'I trust dairy farmers to behave appropriately with it comes to the environment'. Further, the council noted that 'trust has been deteriorating since 2017'. Despite the millions of euros spent on programmes such as Origin Green and the huge political capital invested in convincing the public that 'sustainable intensification' was not an oxymoron, many people are simply not convinced.

The industry knows it. 'The public is feeling that the national value and economic contribution are not as strong as the price being paid on the environmental cost,' the then NDC chief Zoe Kavanagh told the industry-sponsored *Dairy Ireland* magazine. The NDC's own solution to the erosion of public trust in dairy farmers was not to suggest reversing the policies and practices behind it but instead to work on spinning a better story 'grounded in science, and not to simply present a pretty, well-packaged story that could be accused of greenwashing'.[49]

Some eighteen months later, the NDC's ad campaign claiming that Irish dairying is the 'most greenhouse gas emissions-efficient production system' was rejected by the Advertising Standards Authority of Ireland because the NDC was unable to present evidence to support its claims, which included the contention that 'meeting growing international demand for dairy by producing it in Ireland is the best way of tackling the global climate change challenge'.[50]

In May 2024, details of an industry initiative titled Project Connect were released under a freedom-of-information query. It transpired that Bord Bia had warned industry stakeholders that the sector's public licence to operate was under pressure, stating that 'failure to communicate what is actually happening has been recognized time and again as a key limiter of the industry'. It proposed that a new organization be set up to counter what it called the 'negative mainstream narrative'.[51]

Members of the stakeholder group are, predictably, a who's who of industry interests, including ABP, Tirlán, Dairygold and Dawn Meats as well as the IFA and ICMSA. The plan is to set up a company with full-time staff to counter the 'negative perception of key environmental indicators'.

According to Bord Bia CEO Jim O'Toole, 'never have we been at such a critical point of mistrust'.[52] The industry persists in the view that there is an elusive combination of words, images or soundbites that will convince the public that the problems of rising emissions and declining water quality and biodiversity are somehow not the real issues.

*

The public has perhaps not completely fallen for industry propaganda, but neither is it nearly as well informed as it needs to be. Fewer than one in three people in Ireland recognize agriculture as our number one source of GHGs, according to the EPA's 'Climate Change in the Irish Mind' survey.[53]

A study published by the ESRI in 2024 invited participants to maintain an online diary of their daily actions, and to rate those they thought contributed most towards their carbon footprint. Astonishingly, only 4 per cent of respondents mentioned their diet as being a significant contributor. Even among this small subset, many thought – incorrectly – that food miles or packaging were the big issues, and only a tiny number of people identified the role of high-emissions foods such as dairy and beef.[54]

The ESRI study revealed large gaps in the public's understanding around food; for example, reluctance to move away from a meat-rich diet was for many people connected to the belief that it would cost them more money. In fact, largely plant-based diets are typically less costly. The institute observed that there were opportunities for 'low-cost policy initiatives to inform the public about cost-effective ways to reduce the carbon footprint of their diet'. Tellingly, the ESRI also noted that the government's 2023 Climate Action Plan contained no provisions whatever to encourage the shift towards more sustainable diets.

This omission is hardly an oversight. The industrial livestock lobby in Ireland exerts a disproportionate influence over even something as seemingly uncontroversial as advising the public on healthy and sustainable dietary choices. Consider the reaction to an innocuous and light-hearted tweet from the EPA in September 2023 suggesting people reduce their meat consumption to become 'healthier, wealthier, and more fabulous'. IFA president Tim Cullinan professed to be personally 'horrified' at this tweet, adding that there was 'huge rage among our members'. Rage, no less.

The EPA backed down in the face of the furore, and deleted the tweet. The staff member responsible for sending out the tweet quit the agency as a direct result of the 'horrible' fallout from the incident.[55] As we have seen time and again, the livestock lobby in Ireland feels emboldened not just to misrepresent basic science, but also to police the right of others to express opinions, no matter how fact-based or uncontroversial, if they don't fit with the sector's own narrative. Interestingly, while the IFA and others strenuously object to the public 'being told what to eat', this objection in no way deters them from fully supporting and endorsing campaigns, both at EU and national level, encouraging consumption of meat and dairy products.

The EPA tweet kerfuffle was reminiscent of an earlier incident when Taoiseach Leo Varadkar told the Dáil that he was cutting down his meat consumption to lighten his carbon footprint.[56] While noting that 'red meat increases instances of cancer and also contributes more to climate change', Varadkar, a medical doctor, scrambled to reassure the Dáil that he wasn't turning vegetarian or, perish the thought, vegan, adding: 'I had a very nice Hereford steak last night.'

*

In Ireland, the Catholic Church has long understood the value of shaping the views of society by influencing the very youngest through its grip on the educational system. This lesson has not been lost on Ireland's livestock industry, which has long had a presence in schools through the school milk programme. This has been ramped up considerably in recent years with a blizzard of quasi-educational materials supplied to schools aimed to deliver very specific messaging.

I looked into this in depth, and found a variety of agri-industry-funded organizations providing what they called 'curriculum-ready' materials and distributing them free of charge, often presented to teachers as learning resources.[57]

One group, Agri Aware, is a charitable trust controlled and funded by a consortium of agricultural industry players, including Ornua, Glanbia, Kerry, Bord Bia and the IFA. Despite being a tool of industry, it describes itself as 'Ireland's independent agri-food educational body'. Agri Aware distributed a series of four sixty-page workbooks under the title 'Dig In' to over 3,200 primary schools. These include lesson plans linked to strands of the English, maths, science and geography curriculum, according to Agri Aware.

In the workbook's section titled 'Healthy Trees, Healthy Air', it explains that: 'The animals on the farm inhale oxygen and exhale carbon dioxide. Animals and humans need oxygen to stay alive and healthy.' The critical omission here, of course, is methane – but a young child is not to know that. A graphic accompanying the section shows what purports to be a circular flow of trees emitting oxygen, which is taken up by cows, who breathe out CO_2, which trees then absorb.

Another industry lobby group that is highly active within Irish schools is the National Dairy Council. Despite the NDC's name, which would lead most people to assume that it is a state-backed organization, this is in fact a 100 per cent farmer-owned marketing and PR vehicle for the dairy industry. Its 'Moo Crew' campaign, aimed at schoolchildren, includes posters, booklets and worksheets. In-class materials for junior and senior infants on the benefits of dairy products include directing teachers to ask pupils to think of new ways they can incorporate 'milk, yogurt and cheese' into their diets.[58]

Moo Crew also urges teachers to 'encourage your pupils to take home the message of the importance of dairy as part of a balanced diet'. In its lesson plan titled 'Dairy and Eating Sustainably', Moo Crew promotes the benefits of buying local, while neglecting to mention that 90 per cent of Irish dairy products are shipped abroad. The NDC material, in common with Agri Aware, entirely fails to mention methane.

What about the claim by the NDC that 'Irish dairy farms have one of the lowest carbon footprints in the world'? Does Ireland's largely grass-fed system really have a unique advantage? This question was examined in detail in a report commissioned by the European Parliament's Research for AGRI Committee.[59] It found that Irish agriculture was, in fact, uniquely carbon-inefficient. We produced the most emissions per euro of agricultural output in the entire EU 28 in the period from 2012 to 2014. More recent data suggests this ratio has improved, but this is mainly a result of the significant rise in agricultural commodity prices since 2022.[60] Either way, what is clear is that Ireland's grass-based system enjoys no decisive advantage at all in terms of producing dairy produce with fewer emissions than elsewhere.[61]

A spokesperson told me the NDC 'engages with teachers on a regular basis, including focus groups, to ensure the resources are appropriate and developed in line with the Irish Primary School Curriculum'. The NDC, they added, 'is aligned with the Department of Agriculture to deliver the EU School Milk Scheme to primary schools in Ireland'.

I had always assumed that materials provided to our schools had to be carefully vetted to ensure they were both accurate and free from commercial influence or bias. I was mistaken. The National Council for Curriculum and Assessment (NCCA) is the body charged with advising the Minister for Education on schooling from early childhood to post-primary. When I asked about these materials being distributed to Irish schools, the NCCA confirmed that it 'has no remit in what materials are sent into schools . . . it is for schools to decide themselves what materials they use to provide children with appropriate learning experiences'.

Thomas Pringle TD brought this up in the Dáil in December 2020, asking the education minister for her views on 'the proliferation of non-curricular publications from Irish and international private interests seeking to target primary schoolchildren and the action she plans to take to protect the integrity of the education system'.

In reply, minister Norma Foley confirmed the laissez-faire approach of the NCCA, stating that her department had no role in approving, commissioning or endorsing content in any private programme delivered in schools.[62]

This blasé attitude has opened the door to private commercial lobby groups, with the overt support of government departments, and in some cases significant EU funding, to flood Irish schools with misleading educational material in support of clear commercial agendas.

Not content with misleading the schoolchildren of Ireland on the subject of greenhouse emissions, the NDC also presents a warped picture of what actually happens on dairy farms.[63] For instance, it poses the rhetorical question 'Calves are separated from their mothers within twenty-four hours of being born – surely this is unnecessarily cruel and wholly unjustifiable?' The response is simply astonishing: 'In reality, the calves are separated from their mothers for their own good – they can be fed and cared for more easily and appropriately and they suffer less risk of disease and death and from that point of view the practice is perfectly justified.' This is an obvious falsehood, just as it would be wrong to suggest that human babies should be snatched from their mothers shortly after birth 'for their own good'. In the wacky world of dairy lobbying, no statement, no matter how outrageously untrue, appears to be off the table.

As we have seen, the livestock sector is making an oversized contribution to climate breakdown. It is also, of course, especially vulnerable to the very impacts it is helping to worsen. As we will explore later, the transition to a sustainable, low-carbon, nature-friendly and food-secure future will be by no means simple. But the debate on how we achieve it at least needs to be grounded in reality.

6. Getting Our Houses in Order

Sometimes, the most telling signs of change are to be found in the things you cannot see. Driving by any of the new housing estates built in recent years, it often takes a moment to figure out what's different: a subtle alteration of the skyline. The once-ubiquitous chimney stacks are absent from new houses. Modern Irish homes no longer have open fires and, more importantly, have no need for them.

Up to the mid-2000s, Irish homes were rated as among the most energy-intensive in the EU. In the two decades since then, the situation has improved dramatically. In 2005, the average Irish home accounted for just over nine tonnes of CO_2 emissions annually. By 2022, that had been cut in half.[1]

This sea change has been driven by a brilliant (if sadly rare) example of the Irish state taking strong policy action on climate change: the radical reform of our building regulations by the Fianna Fáil–Green government that was in power from 2007 to 2011. Between 2005 and 2009, just 1 per cent of new builds in Ireland achieved an A rating for energy efficiency. By 2021, thanks to the new regulations, the figure was 98 per cent.[2] Almost all recent Irish new-build houses and apartment buildings are fitted with electric heat pumps, allowing the complete phasing out of fossil fuels.

An altogether tougher challenge is posed by the existing stock of a million or so older Irish homes, many of which are poorly built and insulated and in need of an energy upgrade.

Since 2015, the SEAI has supported upgrades to over a quarter of a million properties, but most of these homes still have energy ratings of B3 or worse.[3]

And despite recent improvements, greenhouse emissions per household in Ireland are still close to double the EU average. Just over half of the total greenhouse emissions from Irish households come from oil-fired boilers, with solid fuels and gas each contributing around a fifth. Given that a quarter of all Ireland's energy use and a tenth of our total emissions occur in our residential sector, home retrofits are a critical component of the overall national decarbonizing effort.

Meanwhile, of course, Ireland has a dramatic housing shortage, leading to stratospheric rents and home prices.

The housing crisis is not a simple matter of too little housing for too many people. Part of the problem is that around two-thirds of Irish housing units are under-occupied.[4] The cultural norm in Ireland is that, after a family's children grow up and move out, the parents remain living in a house that may have three or four bedrooms. Another area where we are an absolute outlier in the EU is that only one in ten of our housing units is an apartment, whereas in Europe around half the population live in flats and apartments.[5] It's normal for Europeans to move from relatively large houses to smaller apartments after their children are reared, but that is rare in Ireland.

While Irish homes have become more energy efficient per square metre, over recent decades they have also become bigger, wiping out some of these efficiency gains.[6] The types of homes being built also matters. The typical Irish detached house uses more than twice the energy of an apartment or terraced house, and around a third more than a semi-detached house. The worst energy offenders of all are detached bungalows, due to the high ratio of external area to usable internal space.[7]

*

The problem – a large stock of energy-inefficient housing – is clear. And so is the solution: home retrofits.

The hard part is that retrofitting doesn't come cheap. The Department of the Environment in 2021 estimated that bringing a typical home to a B2 standard, including fitting a heat pump, would cost up to €66,000, depending on the size of the house and how much work was required.

There is a package of grants available that should in theory trim the final cost to the homeowner to around €30,000.[8] But even so, few people have that kind of cash lying around. In 2024 the government unveiled a €500-million low-interest loan scheme, the first of its kind to be backed by the European Investment Bank.[9] This allows homeowners to borrow from €5,000 to €75,000, typically at 3–4 per cent interest, which is well below what banks are charging for standard personal loans. Another advantage of carrying out a home energy upgrade is that homes with a B3 or better energy rating are eligible for 'green' mortgages from the major lenders. These have interest rates typically around 1.5 per cent below the standard rate, which translates into big savings over the lifetime of a mortgage.

Of course, energy upgrading also means direct savings in monthly energy bills, as well as increased levels of comfort and avoiding the health risks associated with damp, poorly insulated homes.[10]

Still, the fact remains that, for too many people at the moment, a home retrofit simply doesn't make sense. One recent case I am familiar with involves a detached larger family home, originally built in the early 1980s, whose owner, an older person, wanted a full energy upgrade, including exterior cladding, solar panels and a heat pump to make the house comfortable for her retirement. The quote for this work came to around €110,000, or €91,000 after grants were taken into account. The wrap-around external insulation alone, after grants, would cost almost €50,000, while VAT accounted for €12,500, or nearly half the value of the total grant aid. This is a huge amount of money for one individual on a fixed income to lay out. If the state seriously wants to tackle carbon emissions in the residential sector and electrify our economy, we are going to have to put a lot more money into grants for retrofitting, and also reconsider whether it really makes sense to be charging VAT for this work.

Another source of funding for retrofits is targeted at people in energy poverty, defined as having to spend more than 10 per cent of household income on energy. Under the Warmer Homes Scheme, almost 6,000 fully funded upgrades of households at risk of energy poverty were completed in 2023 under various SEAI schemes.[11]

Many of our poorest performing houses are owned by older people, for whom the benefits of a retrofit within their own lifetime are far outweighed by the upfront costs. If you are on a fixed income such as a pension, the appetite for taking out a large personal loan, even at preferential interest rates, is likely to be non-existent.

Paul Deane of UCC suggests one solution is property-linked finance. Say you need to borrow €50,000 for a deep retrofit. In this model, the state gives you the full amount, not as a loan but as a charge on the property, so it is paid back in modest monthly amounts over the lifetime of the *property*, not of the owner. If you sell, die or pass on your property, the charge travels with it, and the new owner continues to make the small monthly payments, which might span a fifty- to eighty-year period. 'I think we need to look at radical financing like that,' Deane told me. He pointed out that this idea is not dissimilar to what happened back in the 1950s and 1960s, when the state took the hit on building council houses, to be occupied by people of limited means who paid modest rents over their own and their children's lifetime rather than having to secure a large mortgage.

*

In 2001 my family moved into a solid but chilly mid-terrace house that was built in the mid-1840s. The heating system was connected to an ageing gas boiler, and the house was single-glazed, with some windows having metal frames, which, as we discovered, are an effective way to make a room feel colder. To warm the house, you blasted the central heating; once it was switched off, the cooling was near-instantaneous. There were no radiators in the kitchen area; the only heat source there was an oil-fired Aga cooker that we ran constantly.

Eight years later, we decided to refurbish and retrofit the house, including adding a small extension. We internally insulated the walls and attic, plus the voids under the floorboards, as well as fitting solar thermal panels for water heating and a water harvesting and purification system (installed in anticipation of domestic water charges that never materialized; it hardly ever worked properly and has since been bypassed). The old boiler was replaced with a more efficient condensing model. At the time, there was almost no state support for energy upgrades, so we took the full hit on costs. Despite the disruption and the expense, it turned out in the long haul that our money had been well spent, partly because our heating bills reduced somewhat but also because the house is now vastly more cosy and comfortable.

More than a decade later, in 2022, we decided to take the next step on the path to decarbonization by replacing the gas boiler with an electric heat pump. Given the hefty €13,000 cost of the pump – more than twice that of a gas boiler – the €6,500 grant support was crucial.

Installing the heat pump improved our overall building energy rating (BER) by a full point, from B2 to A2: not bad for a house approaching 180 years of age. Some early teething trouble with noise was eventually solved by the manufacturers, and the system works extremely well: the house maintains a constant comfortable temperature around the clock.

Heat pumps have the potential to break our heavy dependence on dirty imported fossil fuels for home heating. In an era of energy profligacy, heat pumps are elegantly thrifty. For every unit of electricity a heat pump consumes, it generates 3.5 to 4 units of heat. In terms of sheer efficiency, no other technology – other perhaps than the bicycle – comes within an ass's roar.[12]

There are currently around 140,000 heat pumps in Ireland, of which 100,000 are in new builds and around 40,000 retrofitted, according to the SEAI. Data from the CSO shows that 95 per cent of all new builds in 2023 were electrically heated, the vast majority using heat pumps. However, the rate of retrofitting of existing stock remains stubbornly low, at fewer than 4,000 heat pumps being grant-aided annually (although some additional installations will have been carried out without SEAI grant support).

The median price of a heat pump is now nearly €15,000, and the grant is still just €6,500. Even after the application is approved, a householder must cover all costs upfront, then apply to the SEAI to draw down the grant. It is hardly a huge surprise that uptake is very modest.

In 2024, around 44,000 grant-supported home energy upgrades were carried out, a 50 per cent increase on two years earlier, while an additional 2,500 free retrofits were undertaken by local authorities. However, only around one in thirteen of these upgrades involved installing a heat pump. It is essential that the state adopt much more generous measures to accelerate heat pump adoption. It should also place a ban on the advertising or promotion and sale of fossil-fuel boilers – including those mischievously claiming to be 'biofuel-ready'. Every new oil or gas boiler installed in the 2020s is liable to lock that household into burning imported climate-destroying fossil fuels into the early 2040s. The gas industry is trying desperately to slow down the transition away from combustion. In the UK, for instance, the industry has engaged PR firms to try to convince the public that heat pumps are costly, noisy, unreliable or 'Soviet-style'.[13]

Don't be fooled by messaging from Gas Networks Ireland (GNI), which claims to be 'working to decarbonize the gas network with renewable gases such as biomethane and green hydrogen'.[14] GNI claims that by 2045, 70 per cent of the gas in its nearly 15,000 kilometres of pipelines will be 'green hydrogen', and the balance will be biomethane. But green hydrogen is far too expensive and technically problematic to ever have a meaningful role in domestic heating. A scientific review of thirty-two independent studies on hydrogen for home heating found it entirely unsuited.[15] 'Pursuing hydrogen and biogas just to preserve the gas network is totally irrational,' Hannah Daly, professor in sustainable energy at UCC, told me.

As we saw in Chapter 3, biomethane is yet another costly and problematic detour on the road to an electrified, decarbonized future. Despite this, the 2024 iteration of the government's Climate Action Plan still implausibly argues that 'the use of zero-emissions biomethane in heating will also be necessary to achieve our targets'.[16]

Cutting emissions from our homes is further complicated by what is known as the rebound effect. People living in badly insulated homes tend to underheat them, in order to save costs on energy. As home insulation improves, they respond by heating their homes to a higher temperature than previously. As a result, according to the European Environment Agency, 'a significant portion of the theoretical energy and CO_2 savings are often not realized'.[17] This phenomenon underlines the overriding imperative to switch away completely from using fossil fuels for home heating as the only sure-fire way to decarbonize our built environment.

*

Another technology that could have a significant role to play in decarbonizing buildings is district heating (DH). This involves transferring heat from a central source via insulated pipes to homes and other buildings. In Denmark, a country similar in population size to Ireland, two-thirds of all households get their heating and hot water via connection to DH schemes, which are largely powered by a combination of renewable energy and waste heat from incineration or industry. It has six large DH systems in major urban areas, plus around 400 smaller systems across the country.[18] Following the oil shocks of the mid-1970s, Denmark made the strategic decision to go all-in on developing DH systems, and this has been enshrined in law since 1979. The situation in Ireland in the mid-2020s is, of course, quite different.

These systems make sense where they are developed in the vicinity of an existing source of reliable heat, such as an incinerator, power station or data centre, that would otherwise be wasted. In theory, DH schemes could be created using the waste heat produced by data centres or power stations, but many of these heat sources are located too far from residential areas to be of practical use. Even so, according to Codema, Dublin's energy agency, four-fifths of residential heat demand in Dublin could in theory be met by DH. This would be no small task, as it's estimated it would require installing a 1,000 kilometre-long network of trenches to move the heat from central sources to buildings, at a cost of between €2.7 and €4 billion.[19] It is highly unlikely such a major city-wide scheme will ever be undertaken, and DH is far more likely to remain a useful niche option in Ireland's urban energy transition.

Ireland's first DH network was initiated in 2023 in Tallaght, where the waste heat from an Amazon data centre is piped to heat South Dublin County Council's offices and the local library, with plans to extend it to provide heat to 133 affordable apartments in the area.[20]

Dublin's Ringsend waste incinerator has been operating since 2017. Hot water is a valuable by-product of the annual combustion of around 600,000 tonnes of residential and industrial waste at the plant. At present, this water has to be cooled before being dumped back into the Liffey, to avoid damaging aquatic ecosystems. While the Ringsend facility supplies a modest 60MW of electricity to the national grid, an even greater amount of energy – around 90MW – is currently lost in the form of hot water. The good news is that this extremely valuable 'waste' product is to be harnessed to heat up to 80,000 homes and commercial premises in the Poolbeg, Ringsend and Dockland areas via the Dublin District Heating Project, and in the process cut CO_2 emissions by around 16,000 tonnes per annum.[21] In October 2024, COWI, a Danish consultancy, was appointed to work with Dublin

City Council on progressing the scheme. Part of the project is to build a tunnel under the River Dodder through which to carry the heat network pipelines. Given that an effectively free source of heat is available locally from the Ringsend incinerator, there is a very good chance this project will in the coming years deliver Ireland's largest district heating system to date.[22]

Where DH comes into its own is in self-contained projects, such as a large university campus. The Belfield district heating system at University College Dublin (UCD) is a case study of efforts to shift public buildings away from fossil-fuel dependence. This system, which heats eleven buildings on campus, was originally powered by turf, then oil, and then gas. A 1MW air-source heat pump is now the chief provider of energy, backed up by gas boilers at peak times. This initiative has led to savings of around 800 tonnes of CO_2 by UCD every year. The HSE-run St Mary's Hospital in the Phoenix Park installed a large roof-mounted 150kW array of solar panels as well as heat pumps and improved insulation. Crucially, the staff were also involved in behavioural changes and improved awareness around energy use. The net result was what the SEAI termed a 'remarkable' reduction in electricity consumption and gas usage, leading to major cost savings, improved comfort for patients and staff, and a reduction of around 600 tonnes a year in CO_2 emissions.[23]

One major drag on retrofitting and decarbonization efforts has been a shortage of qualified workers. An ESRI study calculated that to meet the annual target of 50,000 retrofits, 15,000 workers would be required.[24] The government provided an additional €117 million in Budget 2023 to upskill and retrain workers for these new positions. We also need to attract workers from overseas with the requisite skills to help fill the many positions opening up in clean energy industries, but the lack of affordable housing makes this much harder. And of course the same limited pool of skilled building workers is in heavy demand for the essential task of building new homes.[25]

*

Just under half a million household properties in Ireland are rented, predominantly in urban areas.[26] The rental sector poses a very specific set of challenges for decarbonization.

The key issue here is the 'split incentive': private landlords bear the cost of energy upgrades and retrofits, but the tenants get the benefit of lower energy bills.[27] The result is poorer energy efficiency, higher energy bills for tenants and limited progress on decarbonization in the rented sector. There are significant incentives for landlords to undertake retrofits, including a tax deduction of up to €10,000 for small-scale landlords who undertake retrofitting with the tenant still in situ. Deeper retrofits are particularly problematic, as they effectively require evicting or rehousing the tenant.

In an extremely tight rental market, steps to compel land-lords to carry out major upgrade work run the real risk of unintended consequences. While the Housing Commission has recommended that all rental properties have a minimum BER in the next five years, Muireann Lynch of the ESRI is sceptical. 'If you were to start passing regulations saying rental properties have to have, say, a minimum B standard BER, you're just going to drive a load of stock out of the market,' she told me. Lynch feels the best way to make rapid progress in the rental sector is to focus on energy upgrades to local authority housing. Irish local authorities and not-for-profit housing associations own more than a quarter of a million social homes, but according to Rory Hearne of May-nooth University, at our current rate of progress, it would take another 120 years to retrofit our entire social housing stock.[28] Obviously the current rate of progress is totally inadequate.

*

One measure that could help offset high energy costs for households while also helping to decarbonize the sector is to fit solar PV panels to just about every building where it is physically possible to do so. In 2022 the requirement for planning permission for homes and other buildings to install solar PV was largely lifted, which dramatically reduces the red tape involved.

The state offers grant support for domestic solar installations, and has eliminated VAT on the supply and installation of panels. For those who have installed rooftop solar PV, another big help was the introduction by the government of the Clean Export Guarantee tariff, which compels utility companies to pay householders for excess solar electricity they supply back to the grid.[29] Domestic solar PV produces the most energy in the middle of the day, when actual electricity consumption at home is usually low, so being able to sell the surplus significantly improves the economics of installing rooftop solar. Combining solar PV with a domestic battery pack means you have the option of storing unused daytime electricity to use at night. It's also possible to top up your battery overnight using cheap off-peak electricity and sell it back to the grid at a small profit the following day, as some users are now doing.

Up until very recently, householders knew little about our domestic energy usage beyond the arrival of the gas or electricity bill every month or two. The meter was usually tucked away in an obscure location and provided only minimal information. Now, with the widespread deployment of smart meters, that is changing. I can log into my electricity provider's website and see exactly how much electricity my heat pump, EV, domestic appliances and lighting are using, as well as being able to plan my electricity usage to avail of lower rates. Our EV, for example, is set to top up most nights at 2 a.m., to make best use of two hours of ultra-low off-peak rates. Similarly, dishwasher and washing-machine cycles can be set to begin after 11 p.m.

While this information is interesting and useful, it does require logging into your online account to access it. I took the additional step of installing a digital energy monitor that visualizes, in real time, household electricity consumption, changing from green to orange to red as consumption rises. If the first step towards energy literacy is awareness, this €100 energy monitor is money well spent; these gizmos should in my view be supplied free of charge by all the energy utilities to their customers.

*

The farm I grew up on was around eight kilometres from Kilkenny city. In those days, that meant we were deep in the countryside. The only buildings along the sleepy country road that passed our gate were farmhouses and the odd cottage, mostly built by the county council and dating back to the 1950s or earlier. We used to walk and cycle this road regularly, and you would rarely be passed by more than one or two cars.

Over the past few decades this quiet country road has come to resemble suburbia. Large detached houses on sites typically of an acre or less now proliferate. There are still no footpaths or street lighting, and you rarely see a cyclist or pedestrian: it's too dangerous, given the volume of traffic on what are still narrow secondary roads. This pattern has been repeated all over rural Ireland in the last half-century, but has really accelerated since 2005, when the Fianna Fáil-led government issued new guidelines to planning authorities on what it euphemistically termed 'sustainable rural housing', but was in reality a free-for-all.[30] Nowhere was safe from one-off housing (defined as a detached house with its own separate septic tank). Even Special Areas of Conservation, Special Protection Areas and National Heritage Areas were, the guidelines said, 'not intended in any way to operate as an inflexible obstacle as such to housing development'.

Gavin Daly of the European Spatial Planning Observatory Network has studied Irish settlement patterns. 'There is this kind of mythology that we've always had this scattered development problem,' he told me. 'We didn't, actually. If you go back to the turn of the twentieth century, most people lived in towns.' Prior to 1970, there were just under 150,000 one-off houses. Since then, that number has almost trebled.[31]

Ireland's unusual settlement pattern creates a 'negative correlation between population density and energy service demands', according to a study by UCC researchers. Large one-off houses have higher energy demands and are too remote to be connected to either the gas network or a district heating scheme. As a consequence, this form of housing is predominantly heated by oil and solid fuels. Ireland's dispersed settlement pattern is extremely difficult to service adequately with public transport and so bakes in heavy car dependence.[32]

The proliferation of one-off houses around the countryside is also directly linked to another thing you notice when travelling around Ireland: the poor state of many of our villages and smaller towns, with empty shop units, boarded-up pubs and abandoned houses defacing the main streets like broken teeth. Half a century ago, these same towns and villages were where the vast majority of non-farmers lived in rural Ireland. These settlements have been hollowed out by the migration to one-off houses in the countryside. Rather than being able to walk or cycle to the local shops or pub, people in one-off houses are more likely to simply bypass their local village and drive to a supermarket in a larger town. Urban dereliction and rural sprawl are inextricably linked.

The contrast between Ireland and elsewhere is stark. Building a one-off rural house in the UK, or almost anywhere else in Europe, without any specific rural-generated economic need is much more restricted than it is here, Gavin Daly told me, 'because they generally recognize that it's absolute madness to have people just living all over the place'. Rural Ireland has suffered from long-term depopulation, meaning there are not enough young people for local schools and sports teams to be viable. Nobody is arguing for the depopulation of rural Ireland, or for preventing building in these areas for people who live and work in them; what is urgently needed is to stop using our rural hinterlands as an unplanned suburban commuter belt sprawling deep into the countryside. Reversing this means revitalizing settlement patterns in rural areas based around villages and towns and freeing people from the traps of social isolation, energy poverty and car dependence.

A hidden climate cost of allowing one-off housing to proliferate is that it has made it far more difficult to find new sites for onshore wind energy, owing to planning rules that prevent the erection of wind turbines within 500 metres of a dwelling. Similar issues have stymied developing routes for new electricity pylons, delaying progress on the North–South interconnector for years.[33]

The government published its draft first revision to the National Planning Framework (NPF) in July 2024.[34] Among its aims is to transition Ireland to a low-carbon, climate-resilient society, and an important element in this transition involves moving away from the current scattergun approach to one-off housing. Ireland's population is expected to have increased by around a million by 2040, so the decisions that are made now in how we plan for this rapid growth will have major ramifications on attempts to decarbonize our economy. The NPF aims to end 'the continual expansion and sprawl of cities and towns out into the countryside, at the expense of town centres and smaller villages'. The target is for at least 40 per cent of all new housing to be delivered within existing cities, towns and villages, on infill and/or brownfield sites. An expert review group on the revised NPF was critical of its lack of ambition to contain or reduce urban sprawl, noting that even if the framework's stated aims were met, that would still allow 60 per cent of new homes to continue to be built as one-offs on greenfield sites.

The 2025 Programme for Government ignored this expert advice and gave its firm promise to 'continue to support one-off rural housing', including financial support for one-off self-builds through the Help to Buy scheme.[35] The fingerprints of the rural independent TDs with whom the government parties are in alliance were all over this section of the programme. When the imperative for what the expert group admitted was the 'politically difficult' option of carefully planned, environmentally sensitive and sustainable development collides with parish-pump clientelist politics and the interests of landowners, in Ireland there is, it seems, only ever one winner.

The aftermath of the devastation wrought by Storm Éowyn in January 2025 revealed the vulnerability of much of Ireland's critical infrastructure to extreme weather, from our 150,000 kilometres of overhead power lines to our water and tele-coms networks.[36] As ESB Networks pointed out, our network length per capita is four times the European average because of our widely dispersed rural population, and this makes the network more difficult and expensive to maintain and repair.

*

Many Irish people who travel to the continent enjoy visiting villages and small towns in countries like France, Spain and Italy. The centres of many of these towns are pedestrianized, with parking around the periphery, making for a calm, more relaxed ambience. The other thing you're likely to notice is that a lot of local people still live 'over the shop', which adds to the sense of human vibrancy.

Returning home, the contrast could hardly be sharper. Many of our villages and towns are so choked with traf-fic, there is room for little else. My home town of Kilkenny some years back took the positive step of limiting traffic on its main street – High Street – by making it one-way. This street is at the heart of Kilkenny's so-called Medieval Mile, a compact city built on a human scale that is eminently walk-able. The council did not, however, take this to its logical conclusion and simply pedestrianize High Street. Consider-ing the huge role that tourism plays in the local economy, the decision seems illogical. I spoke at an event in the town hall recently and afterwards asked a senior council official why they chose a one-way traffic option over simple pedestrianiz-ation. 'Because then I'd be run out of my job,' was his terse reply. The business community in Kilkenny, as elsewhere in Ireland, often seem to conflate the sight of cars driving past their premises with a thriving locality, and this is proving a stubbornly difficult belief to overturn.

There are many settlements around Ireland that could

again become thriving local and regional hubs. The high rates of vacant and derelict buildings in our towns and villages are both a problem and an opportunity. Nationally, of the 166,000 vacant properties recorded in Census 2022, around 48,000 had been vacant for at least six years.[37] From a climate point of view, refurbishment rather than building from scratch is in most cases the best route.[38] There are now a range of support grants available, notably the Vacant Property Grant, which pays out up to €70,000 for qualifying vacant or derelict properties that have been unoccupied for at least two years and were built prior to 2007.[39]

One way to help address the growing need for housing while reducing the climate impacts and revitalizing rural Ireland would be to focus the refurbishment effort on the up to 90,000 properties in Ireland's towns and villages in need of restoration or upgrading. After all, the critical infrastructure for water, electricity and sewerage is already in place, plus there are shops, pubs, schools and other amenities on the doorstep. That's the case made by architect Valerie Mulvin.[40] She hosted an exhibition in mid-2024 which focused on the dereliction in many beautiful Irish towns, like Youghal, Templemore, Clones and Dungarvan. Of these, only Dungarvan – which has been revitalized in recent years as a result of the Waterford Greenway bringing an influx of cyclists – has a population greater than it was in 1840. Dungarvan is just a few kilometres down the coast from Clonea Strand, a place we spent many childhood summer holidays and to where I faithfully return at least once a year. The construction of the Greenway, which hugs some of Ireland's most beautiful coastline, has rejuvenated the whole area, with Dungarvan in particular a major beneficiary of what is now considered to be the jewel in the crown of tourism in Co. Waterford.[41]

Imagination and creativity will be needed to bring Ireland's neglected towns back to their former glories and make them places where people want to live. Mulvin points to the need for local authorities to work with banks and insurers to overcome obstacles to redevelopment.[42] Some years ago, friends of mine were enthusiastic about buying a unit over a shop in the main street in Dún Laoghaire to convert into an apartment as their first home. Despite their best efforts, the planning obstacles meant it became too costly and they eventually gave up in frustration and bought a regular house. The unit, along with scores of others, is still empty today, despite the chronic shortage of housing close to amenities.

Our business moved to the main street in Dún Laoghaire in 1994. Directly across from our office was a jeweller's. It closed down later that year, and the unit, along with the two to either side, has remained empty ever since, a situation local councillor Lorraine Hall described as 'a blight on our town centre'. In the intervening decades these three units have never once been made available to rent or buy.

Laying the foundations for a low-carbon housing sector that is also more resilient to future climate shocks while coping with a fast-expanding population is a formidable challenge. We need to get our planning system into line with European norms and crack down on one-off rural housing for people who are commuting to work elsewhere. We also need a use-it-or-lose-it law on vacant and derelict properties to breathe life back into our city centres, towns and villages. And we need to dramatically increase the financial support to enable homeowners to carry out expensive retrofitting and insulation work that will reap benefits for generations into the future.

7. A Climate of Doubt and Disbelief

When I started working as a climate journalist, I found the key facts around the global climate and biodiversity crisis fairly straightforward to grasp, even without a background in science. I assumed – wrongly – that others in the media would be on a similar learning curve, and that the crisis would inevitably rise towards the very top of the news agenda.

This still hasn't happened. Typical European media coverage of climate change in the early 2000s was three times greater than in Ireland, and there is little evidence of that gap having since closed. A 2020 study found that the climate accounted for well under 1 per cent of total Irish media news coverage.[1] Coverage of major flooding events over the last two decades has focused heavily on local impacts and concerns, with little or no linkage to wider global causes fuelling such extreme weather events.[2]

A former editor of *Fortune* magazine, Eric Pooley, carried out a thought experiment in which he imagined that our leading scientists had discovered a planet-killing meteor was hurtling towards Earth and governments had less than ten years to divert or destroy it. How, he wondered, would news organizations cover this story?[3] He argued that the media would 'throw teams of reporters at it': the race to avert catastrophe 'would be the story of the century'. Pooley expressed deep frustration that, despite the overwhelming scientific evidence of climate impacts capable of bringing human civilization to an equally abrupt halt, the story of the incoming man-made meteor that is climate change rarely makes the front pages or the evening news bulletins.

If this analogy sounds familiar, it's because it became the 2021 film *Don't Look Up*, starring Leonardo DiCaprio and Jennifer Lawrence as scientists who are sidelined by politicians and vilified by the trivia-addled media as they vainly try to warn of an impending cataclysm. Director Adam McKay called his film 'the most thinly disguised metaphor in the history of metaphors'.[4] The comet – spoiler alert – is a cosmic proxy for climate collapse.

Climate, for the most part, simply doesn't fit into the classic paradigm of what constitutes 'news'. The crisis has been brewing for decades, and lacks any finite beginning or clear conclusion. When climate issues have been covered in the Irish media, far too often they have been framed as a 'debate', with more focus on conflict than on the substantive issues. In our broadcast media, this has typically played out as follows: a news item on a climate conference or report briefly outlines the report's key points, before immediately pivoting to an interview with a lobby group spokesperson, to undermine the main points of the report, often by presenting them as an attack on 'rural Ireland'. This in turn helps to set the conflict frame for subsequent coverage, in which the actual science is glossed over.

In the past few years, I have noticed more sustained focus on the science and somewhat less airtime being provided to lobbyists and naysayers. Still, the media remain fond of conflict framing.

*

One of my earliest brushes with outright climate-science denial occurred as a result of a column I wrote on the climate impacts of cheap airline travel in July 2008. Ryanair chief Michael O'Leary took to the *Irish Times* letters page to denounce me as an 'eco loonie', 'eco twit', 'eco clown' and 'eco nut' spouting 'eco babble'. He accused me of publishing 'nonsense, false claims and fictional statistics', none of which he even attempted to refute.[5] The letter was so comically deranged that my wife later presented me with a framed copy of it as a birthday present.

The following year, my weekly column also set me on a collision course with RTÉ's top broadcaster, Pat Kenny, who at that time presented both the *Late Late Show* and a daily radio programme. It began in February 2009, when I penned a piece critical of comments made on the *Late Late* by a long-retired botanist called David Bellamy.[6] Some years earlier, Bellamy had stated he was 'convinced that the continued emission of CO_2 at current rates could result in dramatic and devastating climate change in all regions of the world'. He had further described those attempting to obstruct efforts to tackle global warming as 'criminals'. However, his views had changed dramatically by the time he was given a platform on the *Late Late*. Among his many incorrect statements on the programme was that sun spots were responsible for any recent rise in global temperatures.

I noted in the piece that Bellamy's anti-science bluster went almost completely unchallenged by the host.[7] Kenny's own dismissive stance on climate science was aired a number of times that year on his radio show, most notably in an interview in November, which he opened by asking rhetorically: 'Is the Earth warming at all?' He then told a nonplussed climate scientist, Wolfgang Knorr, that the extra CO_2 in the air was something 'I wouldn't be worried about myself.' Later, he compared the scientific consensus on climate change to Flat Earth theory.[8]

In preparation for the column I was writing, I emailed Kenny a list of questions, seeking to clarify his position on climate science. Instead of replying in writing, he phoned me out of the blue at 9 a.m. one day, an hour before his radio show went on air. An extremely heated exchange, which he insisted was 'off the record', ensued for almost the full hour. His objective, at least as I saw it, was to put a journalistic parvenu firmly in his place. A few days later, my column was published in the *Irish Times*, headed 'Kenny Stirs Up Bogus Climate Change Debate'. I concluded the piece by stating that the 'debate' around climate change was long over and that Kenny had, 'perhaps inadvertently, become a high-profile tool in the hands of the climate denial lobby'.[9]

Around two weeks later, I was invited onto Kenny's morning radio show on RTÉ, ostensibly to have a debate with an Australian mining-industry geologist and well-known climate-change denier, Ian Plimer. Among a host of bogus statements in his book *Heaven and Earth*, Plimer claimed to have tested the hypothesis that increased atmospheric CO_2 creates global warming and found it invalid. But Plimer was not to be the one on trial that day. Even though I sensed I was walking into a trap, I went in anyhow. Sure enough, Plimer accused me of being a 'rent-seeking' hack who made his living out of 'frightening your audience witless', while Kenny joined the attack, stating: 'You are one of those who benefits from the climate change debate.' So, according to Kenny, who was paid around €1 million by RTÉ in 2008, if a journalist writes about climate change, he's just in it for the money.

In the course of a segment that ran for almost half an

hour, at no time did Ireland's foremost broadcaster interrogate either Plimer's credentials to comment on climate
science or his extensive financial links to the Australian
mining industry.[10] Kenny decided instead to make it about
my supposed financial interest in 'pushing' climate change.
(For the record, my total journalistic income in 2009 came
to under €20,000, or the equivalent of one week's salary
for Kenny; and the hundreds of hours I spent researching
and writing my column that year cost me heavily in income
forgone from my altogether more lucrative 'day job' of managing a business.)

In September 2011 I attended a lecture in Dublin given
by Richard Somerville of the University of California. He
told the meeting he was 'still smarting' after a radio interview
earlier that day, in which Pat Kenny had bombarded him
with myth after myth around climate change, each of which
Somerville patiently debunked.[11] When Kenny mentioned
one or two typographical errors detected in the IPCC's
fourth Assessment Report (AR4), Somerville replied saltily:
'A couple of errors were found in a document that is 3,000
pages long; there are errors in the telephone book for Dublin
but that doesn't mean you should abolish the phone company.' Somerville specializes in debunking climate-science
denial, which is rife in the US media, so he schooled Kenny
far more effectively than I had been able to.

It would be more than a decade before I was to share a
studio with Kenny again, this time for his Newstalk radio
programme. Our on-air exchanges in more recent years have
been cordial and professional, albeit with a certain chill in
the studio air. Kenny has pivoted away from outright denial
of climate science and has more recently employed subtler
approaches, such as querying whether there is any point in

a country as small as Ireland acting on climate. In March 2023, for example, he put it to me on air that, with China still commissioning new coal-fired power stations, whatever Ireland does on climate, 'it's such a tiny percentage of the effort that's required that it doesn't really matter whether we do it or not'.[12]

*

As the national broadcaster, RTÉ has long dominated our domestic media landscape, with greater revenue, influence and reach than any other media organization and a unique role in agenda-setting. Put simply, if it's not on RTÉ, it's not considered important. The national broadcaster's near-total abdication of its responsibility to cover the climate emergency throughout the 2010s has to be seen as having played a huge part in the country's desperately inadequate response to the crisis during those years. RTÉ's lost decade of climate coverage, in its own way, mirrored a lost decade for Ireland.

In 2011, RTÉ's environment correspondent, Paul Cunningham, moved to another role in the station and the position was left vacant – or 'suppressed', as a spokesperson told me at the time – for the next five years. That meant, incredibly, that our national broadcaster, with some 2,000 staff, a public-service remit and an annual budget approaching a third of a billion euros, did not employ a single journalist to cover the most consequential unfolding story of the twenty-first century.

That November, the COP17 climate conference was held in Durban. On the opening day of the conference, not only did RTÉ have no reporter in South Africa to cover this landmark event, the conference didn't even merit a mention on that evening's TV news bulletins.

It wasn't until 2016 that the environment role was revived,

and even then it was as part of a newly formulated 'agriculture and environment' correspondent role assigned to George Lee. Two years later, the role was again reconfigured, this time to 'science and environment'. The outbreak of the Covid-19 pandemic in early 2020 saw Lee reassigned to full-time Covid coverage, meaning that RTÉ effectively had no one covering the crucial climate and environment beat for yet another two years.

In 2014, research commissioned by RTÉ's own Audience Council found that climate reporting and debate were marginalized by RTÉ. In a sixteen-month period from January 2012 to April 2013, the study identified thirty major national and international climate stories that were simply not covered by RTÉ. Often, when a climate-related story was covered, it was presented as part of a conflict, such as the *Prime Time* segment titled 'Rural Concern over Wind Farms'. Overall, the study found a general disconnect from and isolation of climate change by the broadcaster from the encroaching environmental disaster. It also noted RTÉ's tendency to treat climate change as an 'international' story and that it 'is almost never presented as a national political issue in Ireland'. The study found that climate change was mentioned in only three of sixty-two national weather event stories, three out of forty-five energy-related reports and just one out of fourteen items relating to conservation or sustainability. Over a two-year period, just one in ten of the 285 news reports on RTÉ's *Six One* that could have mentioned climate change actually did.[13]

In early 2019, it looked like a corner had truly been turned in terms of media and climate. The emergence of the enigmatic Swedish teenager Greta Thunberg and her 'School Strike' movement seemed to inject a much-needed sense of urgency into climate coverage. That November, RTÉ's first – and last – 'Climate Week' was broadcast.[14] It was an impressive effort, with input across radio, television and online, featuring multiple news programmes and documentaries. While at the time it looked like the start of a new era of intense media engagement on the climate emergency, in hindsight it looks like little more than a brief jaunt on the bandwagon.

In May 2021, via a Twitter exchange, I asked the then managing director of RTÉ News, Jon Williams, if it was 'asking too much' for the station to have an environment correspondent. He replied, 'Sadly it is,' adding that RTÉ could not afford to fill this post until everyone paid their TV licence fee.

In July 2021, as temperatures soared and climate-fuelled disasters unfolded across the world, the charity Irish Doctors for the Environment wrote to RTÉ to protest at its climate coverage. 'To report on these record-breaking weather events without mentioning climate change is as egregious as reporting on the unprecedented spike in ICU admissions last April without mentioning a global pandemic . . . it represents an abject failure of journalism and public service broadcasting.'[15] IPCC lead author Peter Thorne also took to social media to express his 'bitter disappointment' at RTÉ's failure, as he put it, to join the dots on extreme weather events and climate.[16]

As the controversy intensified, Williams did something that is almost unheard of at the highest levels of the media: he apologized.[17] 'We were wrong not to make clear connection between recent extreme weather events and climate change,' he wrote, adding that it was a 'sin of omission and reported in good faith. But truth matters. So when we get it wrong, we should say so. Lesson learned. Work to do.' This tweet accompanied an article by Williams on the RTÉ website accepting that attribution science had made it clear that climate change is indeed making extreme weather worse, a fact 'we should regularly remind our audience of'.[18] For the future, he said, 'we will double down on our coverage of climate issues', adding that its public-service remit meant RTÉ had a 'responsibility to lead the conversation about the climate crisis'. He concluded by noting that, from September 2021, every journalist in RTÉ News would be taking part in a workshop on climate science and how to report it.

Williams stepped down from his role as head of news and current affairs the following year.

*

Over the years, RTÉ's flagship current affairs programme *Prime Time* has produced some of Ireland's best investigative journalism. However, tracking the programme closely over the course of the decade 2010–2019, I could identify only two clear instances when it directly addressed climate change.

The first time was in March 2014 and it was, frankly, an unmitigated disaster. *Prime Time* originally planned to have a panel of four. Of these, two would represent the fringe view that climate change is not a major crisis, and a third was the president of the IFA. On the other hand, the overwhelming scientific consensus was to be represented by one person – John Sweeney. In protest at the composition of the panel, Sweeney declined to participate.

Prime Time then removed one of the panellists, the IFA president. This still left the director of a London-based climate-science denial organization, the Global Warming Policy Foundation (GWPF), and Ray Bates, a long-retired adjunct professor of meteorology and member of the Royal

Irish Academy who had emerged in retirement as a hobby climatologist and outspoken critic of mainstream climate science. Defending the programme in a subsequent newspaper article, the show's editor, Donogh Diamond, strongly backed his own decision-making: 'It would be a strange editor who, through his or her own research, and with the assistance of a very talented editorial team, couldn't make a reasonable judgement on the state of scientific knowledge in relation to this vitally important global issue.' In the same article, he referred to Bates as 'one of the country's leading climate scientists'.[19] But Bates never was a climate scientist; he was a retired meteorologist with expertise in the separate field of atmospheric dynamics.

Prime Time revisited climate change in early December 2015, just as the crucial Paris COP21 climate conference was getting underway, with a segment titled 'How Much Will Fighting Climate Change Cost Ireland?'[20] This framed climate action as a threat to state plans to expand agriculture. As presenter Miriam O'Callaghan put it: 'Should we fight our corner or fight climate change?' On the cusp of what was to be a huge expansion in dairy output, Teagasc researcher Gary Lanigan told the programme: 'We will increase production and we do predict that emissions will flatline, and that's not an inconsiderable achievement in itself.' As we have seen, Lanigan's optimism around dairy emissions was misplaced. I also contributed to that programme, though not as a panellist. Inexplicably, the studio discussion that followed the video report yet again placed Ray Bates as the sole scientist on the panel, despite his obvious lack of qualifications. If this really represented what Diamond called the 'best judgement' of RTÉ's flagship current affairs programme, it suggests that something was seriously awry.

Miriam O'Callaghan put it to Bates that 'you are in a tiny minority of a tiny minority' in terms of his insouciance in the face of a global emergency. This begged the question: since it was known within RTÉ that Bates was an absolute outlier on the science, why was he yet again given the sole 'mainstream science' seat on a *Prime Time* panel? Another panellist, Green Party leader Eamon Ryan, was left to state the obvious: 'This debate we're having with Ray isn't happening anywhere else.'

Having said that, *Prime Time*'s coverage of climate change has improved immeasurably since around 2020. In February 2022, it ran a one-hour climate special. 'Future generations will judge us by how we deal with climate change,' is how Fran McNulty introduced the programme. Miriam O'Callaghan added that 'we all know it [climate change] is the biggest crisis facing our planet'. This time, there was no 'maverick' retired scientist on hand to reassure us that business as usual was OK.

In February 2024, *Prime Time* carried an in-depth report on the aftermath of Storm Babet in October 2023, and included expert input from Ben Clarke of World Weather Attribution, climatologist Peter Thorne, and Ciara Ryan of Met Éireann. The unambiguous link between damaging extreme weather events and climate change was front and centre in the report, again with no 'balancing' contrarian voices included. The *Prime Time* report was in turn covered by RTÉ's news bulletins, which reported that the severe flooding in Midleton and elsewhere was made more likely and more severe as a result of 'climate change due to human activity'.[21]

Better late than never.

*

Another key Irish institution that struggled to come to terms with the climate emergency is the ESRI. Of course, its difficulties were not restricted to the area of climate. Its infamous 'Medium-Term Review 2008–2015', published in May 2008, confidently projected annual economic growth rates ranging from 3.5 per cent to 4.1 per cent, all the way to 2020.[22] Dozens of charts, graphs and tables worked out the factors that would drive this sustained level of growth. Incredibly, this report was issued *after* the severe liquidity crisis in the US housing market had already come to light. The writing was already on the wall for an economy carrying a vast amount of housing debt, but nobody in the ESRI – the state's economics intelligence unit – was able or willing to read it. We all know what happened next.

The two most senior authors of this ill-starred report were John FitzGerald and Richard Tol. In February 2015, FitzGerald (who had retired from the ESRI the previous year) told the Oireachtas Banking Inquiry: 'I failed to foresee the impending financial collapse . . . we made a call that Ireland would probably escape it and we were totally wrong.'[23] FitzGerald deserves credit for fessing up; most economists in Ireland also got this horribly wrong, but few ever publicly admitted it.

In 2009, his confidence evidently unshaken by the unfolding financial collapse that he had utterly failed to foresee, Tol authored a short paper titled 'Why Worry About Climate Change?' In this paper – which is still available today on the ESRI website – Tol sees little or no reason for either concern or action. 'Just because something is new and different does not make it wrong. Climate change will take us into uncharted territory, but so do many other things.'[24] Quite.

Tol's overall conclusion, based on what he called 'insights from the economic literature on climate change', is that the impact of climate change is 'relatively small'. In monetary terms, over the entire twenty-first century its total negative impact would be comparable to losing just one or two years' growth. In other words, almost too small to measure. He went on to argue nonsensically that climate change is 'likely to have a positive impact in the first half of the twenty-first century'.

It got me thinking that if Tol and his colleagues at the ESRI, with only the Irish economy to focus on, could get it so horribly wrong about the immediate future, how wise is it to be taking guidance from the same economists on matters clearly beyond their direct expertise – physical sciences – in timescales running into multiples of decades?

Tol joined the 'Academic Advisory Board' of the GWPF, the London-based climate-science denial think tank, in 2010. He left the ESRI two years later to take up a post in Sussex University. In 2014, Tol was one of only two Co-ordinating Lead Authors on Working Group 2 of the IPCC's fifth Assessment Report (AR5), so his views were anything but irrelevant.

Tol quit the IPCC process when his efforts to min-imize the likely economic damage from climate change were rejected by colleagues. In the final version of the IPCC report on 'Impacts, Adaptation and Vulnerability', published in October 2014, Tol's claim that some global warming could be beneficial was removed, as it was found to be based on faulty data.[25]

By that point, Tol's erroneous research had been widely repeated and amplified by climate-change deniers, including in a cover story in the *Spectator* magazine by Matt Ridley, Tory peer, coal-mine owner and fellow member of the GWPF's Academic Advisory Council. This article repeated Tol's unlikely claim that climate change would be beneficial up to 2.2°C of warming and, Ridley added, 'the overall effect is positive today – and likely to stay positive until around 2080'.[26]

*

Richard Tol is an outlier in the field of economics. But the sad truth is that a narrow and deeply misleading approach to the economics of climate change is widespread, and that it has taken its lead from one of the most influential figures in the history of the discipline.

Anyone who studied introductory economics at university during the past few decades was probably assigned the textbook *Economics*, co-authored by William Nordhaus. Nordhaus is widely considered by many to be the first economist of note to try to work out the cost of climate change, and in 2018 he was co-recipient of the Nobel Prize in economics for his work on integrating climate change into long-run macroeconomic analysis.

The climate modelling developed by Nordhaus, known as DICE (Dynamic Integrated Climate & Economics) is at the heart of the economics of climate change embedded in the influential IPCC Assessment Reports, and many other organizations, from banks and insurance companies to the US Environmental Protection Agency, base their climate planning to a large degree around the work of Nordhaus and his acolytes.

According to the DICE integrated-assessment model, at an average global temperature rise of between 2.7 and 3.5°C the global economy achieves what he terms 'optimal' adaptation. To intervene sooner to avoid such temperature rises would, he argues, cost too much money, and besides, the world can adapt fairly comfortably to these average temperature rises with better infrastructure.

This would all be fine were it not for the fact that it is absolute lunacy.

The entire edifice of 'climate economics' inspired by Nordhaus and his DICE model and embedded in the IPCC reports is, in the words of former World Bank chief economist Joseph Stiglitz, 'wildly wrong'. A 2023 study by the UK Institute and Faculty of Actuaries delivered damning findings on the degree to which governments, investors, insurance companies, central banks and even the IPCC were depending on deeply flawed economic models that consistently understated the economic damage of climate change.[27] 'What economists have done is say that climate change is a cat in the bush, not a tiger,' said report author Sandy Trust.

Nordhaus's modelling system ignores tipping points and fails to show significant GDP losses even at an apocalyptic 5°C of global warming. A 2010 paper tested the extreme outer limits of the DICE model by using it to extrapolate the likely impact on GDP of an unimaginable 19°C increase in surface temperature.[28] While this lethal level of heating would extinguish almost every trace of life on Earth, the DICE model was still showing that global GDP would only be halved.

Writing shortly after Nordhaus was awarded his Nobel Prize, John FitzGerald – chair of Ireland's Climate Change Advisory Council at the time – commended Nordhaus for 'integrating the insights from climate science with that of economics, helping us to understand the damage done by climate change and how best policymakers should respond to this threat to humanity'.[29]

There are, it needs to be said, many economists who have not been convinced by Nordhaus's obtuse approach. In early 2024, a group of more than 200 of them wrote an open letter to the European Commission urging it to review and amend its economic forecasting to fully integrate environmental factors into the modelling.[30] They pointed out that assumptions baked into many standard economic models simply ignore environmental and ecological factors.

On the global scale, the wilful misunderstanding and downplaying of climate and ecological risk by economists and their hugely influential role in shaping the public discourse and blunting political response may yet turn out to be the most consequential – and lethal – miscalculation in human history.

*

The one Irish institution you would expect to offer expert guidance and assessment on our climate is Met Éireann, yet for a long time it was at pains to avoid ascribing a climate component to Ireland's fast-changing weather patterns.

Met Éireann has been a member of the World Meteorological Organization (WMO) since 1950, so it might be assumed it would take its lead from the WMO on global climate issues. A decade ago, the WMO was able to write: 'Scientific assessments have found that many extreme events in the 2011–15 period, especially those relating to extreme high temperatures, have had their probabilities over a particular time period substantially increased as a result of human-induced climate change – by a factor of ten or more in some cases.'[31] Yet until very recently, when Met Éireann staff were asked about climate attribution, the standard response was to kick for touch and talk about weather variability instead.

Met Éireann's reluctance to attribute the rise in extreme weather events to global warming was highlighted in the 2019 Oireachtas joint committee report on climate action,

146

in which the committee 'encourages Met Éireann to take a stronger role as the trusted source in weather forecasting to, where scientifically appropriate, link the changes we see to climate change and fully reflect the scientific consensus represented by the IPCC reports in their communications around climate change'.[32] There was no mistaking that this was a slap on the knuckles for the state meteorology service.

Thereafter, there were signs of improvement. Met Éireann forecaster Joanna Donnelly, in an RTÉ interview in September 2021, said bluntly that we are in a 'global code-red emergency situation, and we need to act now'.

However, the message still didn't seem to be getting through in some quarters. A few months later, in April 2022, meteorologist Gerry Murphy in an RTÉ radio interview insisted that 'climate change is very hard to pin down to any specific event', adding that you 'very much have to look at the longer term'. This is technically true, but in the light of clear WMO guidance, as well as the copious climate-attribution studies that are now available to assess the climate-change component of individual recent extreme weather events, there is no excuse for meteorologists to continue to beat around the bush. Murphy's palpable discomfort in discussing the climate was noted by the presenter, Claire Byrne, who steered the interview back to what she called the 'terra firma' of local weather.

Weather forecasters are among the most familiar and most trusted faces on our television screens, and given that weather and climate are two sides of the same coin, broadcast meteorologists are ideally placed to help communicate climate issues to the public. Some, such as US weatherman Jeff Berardelli, are already doing this, by incorporating detailed climate science segments as part of routine weather reports, thus helping the public to better understand the link between a given weather event and the underlying climate signal.[33] Met Éireann should step forward and show real leadership, by embedding discussion of climate-change within its broadcasts.

*

The link between climate and weather is apparently well understood by the industries that are fuelling the crisis. In 2025, RTÉ radio's hourly weather bulletins are sponsored by Grant, a company marketing condensing oil boilers. Meanwhile, Newstalk radio's weather bulletins are sponsored by Ryanair. RTÉ's weather forecast on its TV bulletins has previously been sponsored by dairy giant Glanbia, while its radio show *Country Wide* was at one point sponsored by *Irish* Farmers Journal, which is associated with the IFA. In 2021, I asked RTÉ about the *Country Wide* deal; it insisted that sponsorship arrangements 'do not impact on the editorial independence of the programme concerned'.

RTÉ journalist Philip Boucher-Hayes commissioned a poll of 1,000 people for his *Hot Mess* series of radio documentaries on climate change, and found them overwhelmingly supportive of advertising restrictions in Ireland. He told me that he also 'tried to start a conversation around which kinds of advertising were inimical to climate action, within RTÉ and Virgin Media'. His formal requests to both stations for interviews to discuss their openness to review their advertising policies were, however, denied.

Climate coverage in the Irish media is significantly stronger today than it used to be, but it is still nowhere near prominent enough. I have had a regular weekly environment slot on the current affairs show *The Last Word* on Today FM for more than four years. The segment is typically ten to fifteen minutes long. Newstalk runs a similar weekly slot, called *Green Scene*. And that, as far as regular environmental coverage in Irish broadcast media goes at the time of writing, is just about it. The long-running and highly regarded TV series *Eco Eye* came to an end in 2023, with producer Marcus Stewart citing lack of funding for environmental media.[34]

If audiences subconsciously rank the significance of a given topic based on how much media airtime it receives, then it wouldn't be unreasonable for them to assume that golf, for example, must be far more important than the climate emergency. Nor has anyone ever seen the need to include in golf coverage 'alternative' voices arguing that it is a hoax or that it should be abolished.

RTÉ's managing director of news, Deirdre McCarthy, told an Oireachtas committee in October 2024 that the broadcaster's research indicates there is 'a huge amount of ambivalence and resistance' to the topic of climate change among the Irish public.[35] Despite this, she insisted that the climate remained a 'key editorial priority' for RTÉ. The question all media outlets that are faced with public ambivalence around the climate need to reflect on is whether and to what extent their own lacklustre coverage is at least partly to blame.

The myth of progress is one of our most powerful touchstones. Because we've overcome setbacks in the past, we think we will somehow painlessly deal with climate and biodiversity collapse. Society is operating under what the *Irish Independent* environment correspondent Caroline O'Doherty described to me as a flawed optimism, 'a presumption that the powers-that-be will sort this out, they'll find solutions'. The consequence of our failure to act on the implications of climate science over past decades means that all easy options are long off the table and all remaining choices seem drastic.

'It isn't a lack of intellectual ability to grasp the issue, this is a lack of imagination, and it's possibly a natural bent towards optimism,' she told me. 'I understand that people are now saying "no, I don't want to think about this", it's too big, too uncertain, too much responsibility. And yeah, too scary.'

For many media organizations, the need to urgently ramp up climate coverage has run into the hard economic reality of dwindling revenues. A 2023 report from Coimisiún na Meán, Ireland's media regulator, found that climate and environment were the subjects the public regarded as most worthy of support should government choose to fund public-interest journalism.[36]

Philip Boucher-Hayes is an outlier within Irish broadcast media as a senior journalist with a roving brief who has developed a high level of expertise in climate reportage. His three-part TV documentary *Rising Tides*, broadcast in early 2024, represented RTÉ's largest ever single investment in climate communications, with a budget of €750,000.[37]

Despite this, Boucher-Hayes told me he believes the days of the big agenda-shifting TV documentary are, in today's fragmented media landscape, largely over. 'At the moment I'm leaning much more in the direction that we need to mainstream this [climate] conversation; there should always be now, in every interviewer's mind, what's the climate dimension or element to this story?' In his current role as presenter of *Country Wide*, he feels it's more effective to weave the climate and environmental dimension into storytelling rather than creating discrete 'environmental' output.

Like O'Doherty, he senses a widespread sense of disbelief or unreality in Ireland about just how serious the situation actually is. 'We in the media, we have left the public thinking that this is going to be mostly benign for Ireland. We're going to be one of the lifeboat countries. And yeah, sure, we'll have to cope with increased prices, and our great grandkids are going to have to cope with getting their feet wet more frequently. But in the short to medium term, by which I mean the rest of this century, we're mostly all right, Jack. And that could not be further from the truth.' Food production, he added, 'is something we completely take for granted and we have absolutely no idea how badly disrupted it's going to be in the short to medium term'.

*

If the Irish media is evolving very slowly in the right direction, the same can't be said of the political system. Part of the problem is structural. Our multi-seat constituency system favours public representatives who focus on local and constituency issues, and provides no incentive to promote politicians tuned into wider national and international concerns.

In 2009, the architect and academic Conor Skehan gave a presentation to the Institute for International and European Affairs in which he referred to 'emerging evidence' that rising CO_2 levels are not causing global temperatures to increase, and pushed the idea that global warming had in fact stopped.[38] Following the general election of 2011, environment minister Phil Hogan hired Skehan as a policy adviser. When pressed by *Irish Times* journalist Frank McDonald about his statements on climate change, Skehan described his views as a 'personal position'.[39]

Irish politicians often speak with forked tongues on climate issues. In September 2014, Taoiseach Enda Kenny addressed the United Nations climate change summit in New York with this stirring rhetoric: 'The hand of the future beckons, the clock is ticking and we have no time to waste.' He warned that 'Global warming is a stark reality that can only be dealt with by a collective global response. We are all interdependent and interconnected – we share a common humanity – and each of us must play our part.'[40] Just a month later, Kenny said that Ireland would be 'screwed' if the EU stuck to what he insisted were outdated criteria for reducing carbon emissions.[41] And in November 2015, Kenny told reporters at the Paris conference that climate change was not a priority for Ireland, describing EU Commission targets to cut agricultural emissions as 'unrealistic'.[42]

From a climate standpoint, the 2024 election was genuinely puzzling. Environmental issues barely featured in the campaign. Climate change garnered a fairly insipid eight minutes in one of the leaders' debates – and that was pretty much it.[43] An exit poll found that more than half the electorate felt the government hadn't done enough on climate change, rising to two-thirds of twenty-five- to thirty-four-year-olds.[44] This in no way translated into voting patterns, however, with the Green Party polling just 3 per cent nationally, and losing eleven of the twelve seats it won in 2020. The fate of the Greens in 2024 was typical of small parties in Irish coalition governments, but the reality is that the party's support was small even at its peak.

The state has never made a concerted, well-funded attempt to communicate the scale and magnitude of the climate challenge. In 2019, I asked the climate action minister, Richard Bruton, if such a move was being considered. In response, I was told the government 'will design a nationwide communications campaign early next year'. This never happened.[45] In mid-2024, a limited national and local radio and digital media advertising campaign with a modest €500,000 budget was rolled out by the Department of Environment and Climate Change under the banner 'Climate Actions Work'.[46]

It is described as a national communications campaign 'to tell the story of community climate actions that are already making a difference'. This is fine, up to a point, but it does not go nearly far enough.

While the state has been slow to take the lead on effective climate communication, the same could not be said of those trying to stymie public understanding of the climate crisis. While outright denial of climate science is now rarely aired in the Irish media, a group styling itself the Irish Climate Science Forum (ICSF) has kept the flame burning. The ICSF was set up by Ray Bates in 2016, and in the years since then it has showcased dozens of climate-change deniers from around the world in lectures and presentations in Dublin and online.

In 2018, the GWPF published Bates's critique of the IPCC's 'Special Report on Global Warming of 1.5 °C'.[47] Gavin Schmidt, director of the NASA Goddard Space Institute, wrote a devastating denouncement of Bates's 'silly pseudo-rebuttals to mainstream science'.[48] The foreword to Bates's ill-starred document was written by Ed Walsh, founding president of the University of Limerick and former chair of Ireland's National Council for Science Technology. In it, Walsh backed Bates's claims of 'a lack of scientific rigour and balance in key aspects of the IPCC report'.

In August 2024, the ICSF published a ten-page document titled 'Irish Agriculture and Climate Change – the Good News!', which flatly denied the role of methane as a powerful greenhouse gas and described CO_2 as 'plant food'. The document stated: 'Far from causing a climate emergency, Carbon Dioxide is greening the planet,' adding that this proved there was 'no scientific basis whatsoever for the curtailment of farming in Ireland'.[49]

*

One area in which long-lasting change can be effected is education. Students starting the Leaving Cert cycle in the 2025/26 school year now have the option to study 'Climate action and sustainable development'. This is a welcome if overdue development, though its introduction has to be tempered with the fact that geography is no longer a compulsory core subject on the Junior Cert curriculum.

Earlier, we saw how agri-industrial interests have been allowed to flood schools with marketing and PR materials. Even when there are not overtly commercial interests at play, the National Council for Curriculum and Assessment's hands-off approach can produce worrying anomalies. Private educational publishing companies operate with minimal state oversight, as I was to discover some years back when my then eleven-year-old daughter pointed out to me that her geography textbook included some statements that even she as a primary-school pupil knew were simply inaccurate. The textbook in question, *Unlocking Geography*, was published by Folens. The section titled 'Global Warming' presented the subject in the form of a 'debate' between two hypothetical scientists on the question of whether or not climate change was even real.

I wrote to the publishers on behalf of An Taisce's Climate Committee to point out the issue.[50] In reply, Folens stated: 'unfortunately neither the Department of Education nor the NCCA are willing to either review or approve educational content for publishers'.[51] Folens did, somewhat reluctantly, accept that the chapter was seriously flawed and needed to be rewritten, which I undertook on a pro bono basis, with John Sweeney checking the revised version to ensure its scientific accuracy.[52] Despite the rewrite, it took several years for Folens to withdraw the original textbook.

The hallmark of Ireland's collective response to the unfolding climate emergency has been complacency, though there are signs that this is changing – albeit slowly. At the same time, with the far right in the ascendancy in the United States and elsewhere, misinformation and disinformation about climate is again on the rise globally, with social media being used to fan the flames. For too long, Irish media and institutions have been pitifully weak in the face of these forces. We are now going to have to radically up our game.

8. A New Food and Land Vision

A fodder crisis in Ireland during the summer of 2013 saw 'widespread hunger among animals and some deaths, as well as the premature slaughter of an estimated 30,000 cattle below their target weights'.[1] Fleets of lorries from the UK and France delivered huge consignments of hay to farms across the country.

Five years later, an exceptionally wet spring was followed by an extended summer drought, resulting in another severe crisis as farmers ran out of feedstuff for their animals. In August of that year, the government announced schemes totalling €7 million to help farmers buy the thousands of tonnes of imported fodder they needed.

In 2022, the government approved a €56-million fodder support scheme to offset higher import costs of fodder arising from the war in Ukraine and from near-drought conditions that summer. The following year saw extreme rainfall across much of the country, beginning in September 2023 and continuing almost continuously until April 2024. This led yet again to drastic fodder shortages by the summer of 2024, with the ICMSA once more calling for taxpayer-funded fodder schemes to ease the pressure.

What is clear is that Ireland's massively oversized livestock sector, which includes over 7 million cattle, is lurching from one crisis to the next. And such crises are only going to become more frequent and more severe. A UCC-based research team that conducted an analysis of the 2018 fodder crisis concluded that climate change would at least double the likelihood of similar severe crises in the decades ahead.[2]

This begs the question: do we have a fodder problem – or an overstocking problem? The myth is that Ireland is a singular grassy paradise. The reality is that Irish cattle farmers regularly struggle with the basic task of keeping their animals fed, and that the problem is only going to get worse.

This is just one of many reasons why we need to completely rethink the way we use the land and practise agriculture in Ireland.

*

Imagine for a moment that Ireland was a blank slate, and you had the task of deciding how best its nearly 7 million hectares of land would be shaped so that we lived within ecological limits, with a thriving, resilient agriculture system operating in harmony with a landscape of intact ecosystems teeming with wildlife, our rivers and lakes running with pristine water and rich in aquatic life.

The first thing this exercise requires is to recognize just how far we are from this idyll. Ireland is now among the most nature-depleted countries in the world.[3] The fact that this is so poorly understood by the general public is at least in part a testament to the sustained efforts of state agencies and commercial vested interests to paint a deeply misleading picture of Ireland's ecological health. Biological diversity and functioning ecosystems are not just about cuddly animals and photogenic landscapes; they are as crucial to human welfare as the very air that we breathe.

Ireland's land-use system is, as we have seen, profoundly imbalanced. The dominance of livestock, and the low level of food production for direct human consumption, explain why the agricultural industry is a disproportionate contributor to Ireland's greenhouse emissions, biodiversity loss and water pollution.

Significant effort and investment have flowed into trimming the ecological and climate hoofprint of our livestock system through largely technical measures, but with very limited success. Far less effort has been directed towards developing and expanding alternative forms of agriculture that are less polluting and more efficient at feeding people.

Horticulture accounts for under 2 per cent of the country's farmland. But land under horticultural production produces income per hectare at three times the national average, and despite its tiny footprint it accounts for one in ten agri-food jobs in Ireland, while producing well over half a billion euros' worth of farmgate value.[4] This is by far our most efficient food system – and yet domestic production of fruit and vegetables has been in steep decline for the past sixty years or so.

Meanwhile, we import large quantities of fruit and veg, from Spain, the Netherlands and the UK in particular. While some of our imports are of products such as bananas and citrus fruits that cannot be grown locally, and others are crops out of season in Ireland, nearly a third of our total imports are of fruit and vegetables that could be grown domestically. Ireland's horticulture sector is heavily dependent on migrant labour, with seasonal workers now coming from as far away as Thailand and the Philippines as well as eastern Europe. According to Bord Bia, there are now just sixty field-vegetable growers left in Ireland – an 85 per cent decline in the last twenty-five years – as most of the smaller producers have left the sector.[5]

At the moment, the rule in the industry is: go big or go bust. One home-grown operation that has managed to scale up successfully is the family-owned Keelings, based in North Co. Dublin. With an annual turnover of around €330 million and 2,000 employees, it grows more than 100 million strawberries for the Irish market in a 50,000-square-metre greenhouse complex, and uses biological crop control to reduce the use of chemical pesticides. It is the closest thing we have to the Dutch model of ecologically sustainable intensive food production for humans.[6]

Horticulture received some rare sustained political focus with the launch in May 2023 of the National Strategy for Horticulture, which was commissioned by Green Party minister of state Pippa Hackett. Its ambition is to grow the sector by a third by 2027, while increasing farmgate value to around €680 million. As Hackett stressed in introducing the strategy, climate impacts now being experienced across Europe, as well as geopolitical events, most notably the Ukraine war, have put the focus firmly on the need for Ireland to reduce its high dependence on imported foodstuffs. Their cost – or even availability – is coming increasingly into question as climatic and political shock waves spread across Europe.

All other considerations aside, horticulture is the most resource- and carbon-efficient way of feeding people. During the Second World War, when Britain was under attack and many of its sea lanes were cut off from imports, it turned to 'victory gardens' to help feed the population. Within three years, there were 1.7 million public allotments and a further 5 million private gardens producing food.[7] The idea spread to the US, where two-fifths of all the fresh vegetables consumed in 1943 came from victory gardens.[8]

During the same period in Ireland, many local authorities rented out vegetable plots to householders. In 1942, for instance, Cork City Council was running some 2,500 allotments.[9] This is as many as the total number of allotments in Ireland today. In stark contrast, Denmark, with a similar-size population to Ireland, now has over 40,000 allotments.

During the two world wars, compulsory tillage schemes were introduced in Ireland in a bid to avoid food shortages, and cereal production for human consumption rose sharply. Cereals comprise around a quarter of the average daily energy intake of an Irish adult, a fifth of their protein intake and nearly half of their daily fibre intake. Less than 7 per cent of Irish farmland is currently devoted to tillage and, as we have seen, most of what is produced – whether barley, wheat or oats – is for livestock fodder.[10] Most of what humans consume, including almost 100 per cent of our milling flour for bread production, is imported.

A narrative routinely repeated by agri lobbyists in Ireland is that the land is best suited for grass growing. However, a Teagasc paper refutes that: 'Ireland has amongst the highest levels of crop productivity, with the highest average wheat yields and the second-highest barley yields in the world.'[11]

I came across a fascinating 1938 paper by E. J. Sheehy, a UCD lecturer in animal nutrition, that still rings true today. 'If it is decided then, as a national policy, to have a prosperous rural community and especially a high population on the land, the nation must be prepared to foot the bill by subsidizing tillage or by adopting measures which will have a similar effect. The problem is a national one, not one to be financed by the farmer,' Sheehy wrote. 'If, therefore, we are to have, as a national asset, a reasonable proportion of tillage and a prosperous rural population, the entire community must bear the burden.'[12]

Whatever the circumstance, one thing is clear: when the chips are down and the prospect of real food shortages loom, people turn to horticulture and tillage to meet their immediate food needs.

*

Whatever kind of agriculture is being undertaken, organic farming is, by some distance, the system most friendly to nature and biodiversity. Across Europe, roughly a tenth of farmland is managed organically, and the trend is distinctly upwards. The EU's Farm to Fork strategy set the ambitious target of a quarter of all farmland being managed organically by 2030. As of 2023, Austria led the EU, with 26 per cent of its agricultural land farmed organically, closely followed by Estonia and Sweden. Despite trading internationally on its 'green' image, organic farming in Ireland has long languished in the doldrums, the country having the third lowest percentage of organically farmed land in the EU in 2023.[13]

Organic agriculture is certainly kinder to nature, but is it really capable of feeding humanity? A major review over a forty-year period concluded that while yields were on average 10 to 20 per cent lower with organics, this system was more profitable for farmers as well as being more environmentally friendly. Profitability improves because input costs, including fertilizers and pesticides, are drastically reduced, and because the public are prepared to pay more for organic produce.[14] The somewhat lower yields of organic farming would be a problem if we simply converted to organics at scale while leaving the land-use mix unchanged. But in Ireland, where our farmland is overwhelmingly dominated by cattle, and where most of our grains and cereals are consumed by animals, we have vast capacity to increase our food production, even with wide adoption of organic systems.

The good news is that we are now moving in the right direction. In 2023 alone, around 2,000 farmers joined the

Organic Farming Scheme, doubling the number of organic farms and bringing the total area farmed organically to around 180,000 hectares, or 4 per cent of our total farmland. The government's Climate Action Plan aims to have 10 per cent of our land farmed organically by 2030. Whether this greening trend in Irish agriculture survives a government in thrall to rural independents remains to be seen.

There are clear climate as well as ecological benefits to organic farming. For livestock, organic systems mean fewer animals per hectare. By eliminating chemical fertilizers, organic systems reduce the powerful greenhouse gas nitrous oxide (N_2O). Another major benefit is in easing the pressure on nature from pesticides. Around 3,000 tonnes of pesticides are used annually in Ireland, the vast majority for agriculture and by local authorities, while 450 tonnes are sold in a completely unregulated way to the public via garden centres and even supermarkets.[15] The agri-chemical industry has annual revenues of around €250 billion, and is a powerful lobby against nature protection and organic agriculture. It has fought hard to block the EU's Farm to Fork ambition to cut pesticide use in half by 2030. Researchers estimate that agri-chemical corporations spent over €40 million lobbying to overturn new EU rules restricting the use of pesticides, using the cover of food security in the wake of the Ukraine invasion to thwart pesticide reforms. Farm pressure group Copa-Cogeca (to which the IFA is affiliated) has an annual lobbying budget of €1.5 million, and has also opposed pesticide regulations.[16]

While devastating to wildlife, including pollinators, widespread pesticide usage also poses a significant threat to human health. Acute pesticide poisoning is a major global health challenge, with many millions made ill, along with tens of thousands of deaths every year directly attributable to exposure to pesticides.[17] Pesticides also pose a contamination risk to drinking water. Uisce Éireann recorded fifty-two cases of pesticide residues above the safe level in drinking water in 2023.[18]

I recall a surreal moment on the *Late Late Show* in 2019 when a young woman came on to discuss her invention, the Safe Scrub Sprayer. As she explained, her father, a farmer, had been out hand-spraying weedkiller. Despite wearing protective gear, he became ill and required medical attention as a result of being exposed to the toxic chemicals. Her solution was to develop a spray unit that could be operated remotely from the tractor cab.[19] While human safety is, of course, vitally important, what struck me was that at no point during the interview did anyone pause to reflect on the consequences for the natural world of the widespread use of these manifestly dangerous chemicals.

While the agri-chemical industry and many farm lobbyists will see the watering down of regulations on pesticides as a win, this is in reality a Pyrrhic victory. Nearly half of the world's insect populations are in sharp decline, with two in five species facing near-term extinction.[20] Why does this matter? Put simply, the untrammelled use of pesticides is killing the natural world. A 2025 study assessed 1,600 species and found one in five pollinators now at risk of extinction, with agriculture and climate change the two key threats. In North America alone, pollinators provide over $15 billion in value to agriculture.[21] The insect kingdom is the foundation of the entire global food chain, underpinning all life on Earth. We are witnessing in one or two generations the chaotic unravelling of aeons of evolutionary complexity, and our industrialized food production systems are the number one driver of this ongoing collapse.

Bird populations, many of which depend directly on insects, are in freefall globally. Populations of flying insects across Europe have declined by around 80 per cent over the last three decades, causing bird populations to decrease by more than 400 million in the same period. In Ireland, two-thirds of our farmland bird species are in decline, with one in four seriously threatened. Habitat conversion to intensive agriculture along with pesticides, fertilizers, invasive species and climate change are the main contributory factors to this unfolding ecological catastrophe.

Projections from a plethora of studies in recent years highlighting the dire consequences of the agri-industrial model of global food production on the natural world have largely fallen on deaf ears. Farming systems that depend on repeatedly dousing the landscape in toxic chemicals, be they pesticides or chemical nitrogen, are the very definition of unsustainable. Remember, what we now call 'organic farming' used to be called simply 'farming'. For centuries, countless generations of farmers understood that their prosperity, indeed their very survival, depended on having the skills to work with the land and to respect rather than trying to eliminate nature. Much of this expertise has been lost in the scramble for profits and productivity gains.

*

There's an old joke that goes something like this: give the Dutch Ireland and they could feed the world; give the Irish Holland and they would drown. It may be a little harsh, but it contains a germ of truth. Holland, a small country, less than twice the size of Munster, yet with a population of 17 million, is, remarkably, the world's second largest food exporter, behind the United States.[22]

While historical hunger is etched deep in the Irish psyche,

for the Dutch the experience of famine is an altogether more recent trauma, and has shaped their attitude to food security ever since. In the brutal winter of 1944–5, adult rations in Amsterdam fell to around 580 calories a day, largely due to the German army blockade on food. Thousands starved to death, and people were reduced to eating tulip bulbs and sugar beet to survive. Many survivors suffered lifelong health impacts.[23]

Spurred at least in part by the folk memory of severe food shortages, in subsequent decades this tiny country has become a global powerhouse of food production and exporting. As the world grapples with the question of how to feed a population likely to approach 10 billion by mid-century, the Dutch agricultural miracle shows just how much can be achieved using limited resources.

At the turn of the twenty-first century, there was growing concern in Holland regarding its ability to feed its large and growing population on such a small land base. From this, a mantra arose: produce twice as much food using half as many resources. Holland is now among the world's largest exporters of agricultural and food technology, and it is a trailblazer in cell-cultured meat, vertical farming, seed technology and a host of innovations that have allowed farmers to cut water consumption as well as greenhouse gas emissions. Incredibly, the Netherlands now accounts for a third of the global trade in vegetable seeds.

The country's extensive network of greenhouses allows controlled growing conditions throughout the year, including in many cases the virtual elimination of the use of chemical pesticides. Companies like Koppert Biological Systems sell farmers biological pest control solutions, including ladybug larvae and nematodes that eat the larvae of flies that attack commercial mushrooms. Koppert also maintains hives of bumblebees, which are leased out to farmers for pollination.[24]

Wageningen University & Research (WUR) is at the heart of the Dutch agricultural transformation, a global leader in agricultural research.[25] It is surrounded by a cluster of research centres that are collectively known as Food Valley. There is simply nothing remotely like this in Ireland.

A key vulnerability of farming under glass is energy: greenhouses consume almost a tenth of the total natural gas used in Holland.[26] But three-quarters of the gas used in Dutch greenhouses fuels combined heat and power (CHP), a system with 90 per cent efficiency. It produces electricity as well as hot water, and the CO_2 it generates is pumped into the greenhouses (in moderate, controlled amounts, CO_2 is in fact a plant fertilizer).

Holland is also Europe's largest meat exporter. In 2020, Holland had 3.8 million cattle, 12 million pigs and 90 million chickens. The previous year, the Dutch Council of State ruled that the Netherlands' system of nitrogen permits was failing to protect Natura 2000 nature reserves. The government promised 'drastic measures', and in December 2021 it announced a €25-billion plan to sharply reduce the total number of livestock. This included paying some farmers to either leave the industry or transition to less polluting form of agriculture. 'We can't be the tiny country that feeds the world if we shit ourselves,' as government MP Tjeerd de Groot memorably put it.[27] Dairying was identified as the number one culprit, yet dairy accounted for only 1 per cent of Dutch GDP. This seemingly innocuous regulatory shift triggered what is known as the *stikstofcrisis*, or nitrogen crisis.[28] It was pounced on as a cause célèbre by right-wing activists around the world. In 2023, Donald Trump praised farmers for 'courageously opposing the climate tyranny of the Dutch government'.[29]

Covid-era conspiracy theories and right-wing political opportunism mixed with legitimate concerns to form a dense miasma of misinformation about the scheme to reduce nitrogen pollution. Many Dutch farmers did have genuine reason to feel aggrieved about how the process was handled, especially given that nitrogen emissions from agriculture had already fallen by nearly two-thirds since 1990. But in the current global political environment, any attempt to mitigate agricultural pollution is likely to be hijacked by right-wing elements seeking a culture war, and so it was in the Netherlands. The *stikstofcrisis* had, by early 2024, seen the far-right Party for Freedom (PVV) form part of a ruling coalition following a surge of support for right-wing parties, including the newly formed Farmer-Citizen Movement (BBB). The episode is a warning from which Ireland ought to learn.

One thing the Netherlands and Ireland have in common is that both are significant exporters of meat and dairy products. There the similarities largely end, however – and in most respects the Netherlands is way ahead of us.

Apart from being vastly more productive, Dutch agriculture is also far more diversified. While our politicians make much play of Ireland 'feeding the world', the Dutch are actually doing it through a combination of innovation, diversification, entrepreneurship, high technology and collaboration between academia and industry. Like all intensive industries, Dutch agriculture has its pitfalls and challenges, but if Irish agriculture wants to look to a model for the future that combines productivity and technology with efficiency, sustainability and resilience in the face of our changing climate, then Holland has much to teach us.

*

The consolidation of power in Ireland's food industry mirrors international trends over the last half century and more. Our most financially successful operations – large dairy farms – might best be described not as farms in the traditional sense of the word, but instead as bulk commodity production units for corporations. Inputs in the form of animal feed and fertilizers arrive onto the farm, whose main (or, in some cases, sole) on-farm crop is fast-growing monocultural ryegrass. A truck arrives every day or two and collects milk to be transported to a manufacturing facility. From there, only around one litre in ten will be sold to the Irish public as liquid milk or other dairy products, such as butter, cheese or cream. The vast majority of the farm's output will be shipped overseas, much of it as powdered milk, an inferior replacement for human breast milk, in emerging markets.

While traditional farming depended on the farmer's intimate understanding of soils, crop rotation and the cycles of predators and prey, much of this skill has been lost as modern farming depends primarily on overpowering the land with chemical pesticides, fertilizers and heavy machinery. We have been led to believe that there is simply no choice, that industrial agriculture is the only possible way of feeding a global population of more than 8 billion people. In reality, industrial agriculture uses around three-quarters of the world's farmland while producing less than a third of the total food eaten by humans. The majority of the world's food needs are met by peasant farmers, even though they own or farm barely a quarter of the world's farmland.[30] Peasant/smallholder farmers in the global south harvest more than half of the world's total crop calories, including 80 per cent of rice and 75 per cent of groundnuts.[31]

Perhaps the heaviest toll taken by industrial farming systems is on topsoil. At least 75 billion tonnes of topsoil are being destroyed globally every year, largely as a result of unsustainable farming practices. This translates conservatively into an annual loss of around 10 million hectares of cropland owing to soil erosion.[32] That's the global equivalent of losing more than twice the total farmland in Ireland every twelve months. Soil is, in human timescales, non-renewable. Once it's destroyed, it's effectively lost for ever.

Some 95 per cent of our food depends on soil. The UN Food and Agriculture Organization has warned that, based on the current rates of degradation, virtually all the world's topsoil could be lost within the next sixty years.[33] As if that weren't enough, soil destruction is also a major global source of greenhouse gas emissions.

Almost a century of data on soil biomass, including the density of soil invertebrates, was reviewed in the UK's first full national assessment of earthworm numbers. It found shocking reductions in earthworm abundance in UK soils, with declines of between 33 and 41 per cent recorded in just the last twenty-five years, notably in farmland and 'likely driven by agricultural intensification'.[34]

Coming to understand that industrial agriculture is not in fact feeding the great majority of the world's population today, and that it is inflicting grave damage to biodiversity along the way, frees us from the misconception that there are no alternatives to this manifestly ecocidal system. The choice the people of Ireland, Europe and beyond now face is whether or not we fatalistically persist in subsidizing a failing and destructive agriculture model. Given the parlous state of global geopolitics, combined with the fast-accelerating climate emergency, we need a radical new vision for both use and stewardship of the land of Ireland, which is our most precious resource.

It is clear we need to change course urgently, but what are the alternatives? French researchers modelled a scenario for putting European agriculture on a sustainable pathway by 2050. It involved phasing out pesticides and synthetic fertilizers, promoting natural grasslands, and extending agroecological infrastructure, including hedgerows, native trees, ponds and stony habitats. Overall, this scenario led to a 35 per cent decline in total agricultural production, yet the food needs of all Europeans were still fully met, and Europe was still able to export surplus cereals, dairy products and wine. In this scenario, overall greenhouse gas emissions from agriculture are down 40 per cent by mid-century, while biodiversity is rebounding and water and quality is stable and improving. The authors point out that a remodelling of our food system along these lines would be no more radical than what occurred across Europe between 1950 and 1980, with the rapid industrialization of the food system.[35]

*

Fergal Anderson is a small-scale organic farmer based in Loughrea, Co. Galway. He is one of the founders of a new grassroots movement known as Talamh Beo ('living ground'), which practise agroecological farming, incorporating ecological and social considerations into agriculture. In 2024 Anderson co-authored a document calling for the creation of a new 'local food producer' status to help boost farmers who are producing and supplying products directly to local businesses and households.[36]

Talamh Beo argues that in order to revitalize rural communities and help them achieve resilience in the face of climate change, the state should look to support small-scale local food producers through income support schemes, labour and finance incentives, projects to improve access to land to new farmers, and support for developing resilient short supply chains. It believes that farmers and communities should be at the centre of decision-making for food and agricultural systems and that they should be an integral part of the development of agricultural policies. Tellingly, while the industrial fertilizer sector is represented in the Teagasc Authority in the person of a ministerial nominee, there is as yet no such representation for small-scale community-led agroecological organizations.[37]

I met Anderson and other Talamh Beo members at the 'Feeding Ourselves' conference in the Cloughjordan eco village in 2024. It was an invigorating, inspiring event, held over several days, with farmers and food producers, ecologists and environmentalists, exchanging experiences and insights. The whole event was about as far from the media caricature of farmers and environmentalists at one another's throats as you could imagine, and it made me wonder to what extent this supposed discord is being deliberately manufactured and exaggerated. Many farmers are also environmentalists, though they might be reluctant to admit as much in some settings. They know well the damage the current system is inflicting on our soils and our landscapes, yet are struggling to find ways of breaking free from a system that is dominated by cattle barons and agri-industrial PLCs. Despite the many problems faced by Irish agriculture, we do have the advantage that, unlike the UK and US for instance, we still have many small farmers and we have not yet seen the kind of land consolidation that is typical of the industrialization of rural areas. For Talamh Beo, the guiding principle is to achieve food sovereignty, which entails wresting back control from global players and producing food for people, not just as a raw material for corporations.

Another Talamh Beo co-founder, Thomas O'Connor, and his wife Claire sold the family pub and bought a twenty-five-acre smallholding near Tralee that combines horticulture, permaculture, agroforestry and native Irish woodland. He takes a bleak view of the near future for humanity, but rather than despair, he believes that now is the time to act and to prepare. 'It's about gathering knowledge and information, building strong communities, and ensuring that as the storm passes, there's something left for the next generation to build on,' he told me.

Supporting the building and nurturing of resilient local food producers really shouldn't need to be seen as an ideological issue. Climate change poses existential threats to the world's food production systems. In some places the main hazard will be extreme heat and/or water shortages. In others, it will be soil degradation. And some of the world's breadbaskets will be particularly vulnerable to extreme weather events that will devastate production.[38] These forces are likely to be extremely disruptive to the ability of a heavily import-dependent country like Ireland to meet its basic food needs. Ireland must build the capacity to fulfil the vast bulk of our nutritional requirements from within our own shores in a system that is resilient to supply-chain shocks as well as to the effects of climate change and extreme weather events here.

*

When the first human settlers reached Ireland around 9,000 or more years ago, they would have found a post-glacial landscape of extensive forest cover interspersed with bogs, dominated by oak, elm and pine trees. The forests teemed with brown bears, wolves, red squirrels and boars. Forest clearance began as far back as 5,000 years ago, with elms in particular disappearing in large numbers, due at least in part to clearance by Neolithic farmers. Climatic changes would also have contributed to the decline of forests, with wetter and cooler conditions leading to the expansion of bogland.[39]

Human impacts over millennia have radically reshaped our landscape, leaving only the faintest traces of its original condition. The Norman invasion in the twelfth century accelerated forest clearances, and by 1600 forests covered around one-fifth of Ireland.[40] The iron-smelting industry used huge amounts of wood to produce charcoal for smelting. The manufacture of wine casks and other wooden barrels, as well as the demand for Irish oak for shipbuilding for the British navy, saw vast swathes of native forest felled in a relatively short period. By the end of the nineteenth century, barely 1 per cent of Ireland was forested.

That figure is now around 11 per cent, largely owing to the expansion of timber plantations over the past thirty years. Most of these plantations are spruce monoculture, of little or no ecological value. They are largely operated by Coillte, the state-owned commercial forestry business, which today manages 440,000 hectares, or around 7 per cent of the total land area of Ireland.[41]

Ireland's managed forests are now a net emitter of CO_2, to the tune of around 420,000 tonnes per year, due in part to high emissions from trees planted on peat soils and to the increasing harvest levels of forestry that will result during this decade and next from the age profile of plantations. It is now universally accepted that we have to rethink our approach to forests and forestry. According to a 2024 paper in the *Irish Forestry Journal* (an industry publication), change is now needed: 'To meet the demands of the twenty-first century, in terms of climate change mitigation and resource supply, forestry must shift its focus towards nurturing ecosystems that are biodiverse and resilient, rather than emphasizing the accelerated growth of trees for commercial harvest.'[42] The Tree Council of Ireland's 'Forestry Programme 2023–2027' set out the goal that the annual share of new planting should be 50 per cent broadleaf. To boost the planting of native trees, the state no longer requires people planting small areas to hold an afforestation licence. The new scheme also supports agroforestry, which is the combination of trees and crops.

Commercial forestry as currently practised in Ireland is vulnerable to the sort of extreme weather events that are becoming increasingly common. Storm Éowyn's toll on Irish forestry is estimated at around €500 million, with as many as 30,000 hectares destroyed, or the equivalent of two and a half years' total harvest. Eighty per cent of the destroyed forestry was fast-growing commercial conifer – which has shallow roots – while only 5 per cent were broadleaf trees.[43]

A new vision for forestry in Ireland would see us pivot sharply away from non-native commercial plantations and towards the restoration of native forests of slower-growing but more sturdy broadleaf trees left intact over many decades; these provide ecological benefits, sequester CO_2, reduce flooding risk by stabilizing the uplands and represent a valuable tourism asset.

*

Our upland areas of hills, bog and mountains are of marginal agricultural value. Largely inaccessible to machinery, these lands, which ought to be of the highest ecological value, are instead largely dead zones, heavily grazed by sheep (and in some cases also by deer and goats), and 'managed' by frequent burning, legal and otherwise.

The grazing of sheep causes huge ecological damage to sensitive upland habitats. In addition, upland areas are routinely burned in order to keep vegetation in grazeable condition. This burning is encouraged by Teagasc, which argues there is a 'strong need to bring upland areas back into active management'.[44] It has devastating consequences. For instance, in early 2022, firefighters were reportedly 'sickened' by the widespread destruction of wildlife in the once-pristine wetland of Na Gorta Dubha in Co. Kerry. 'It was a mad act, especially in terms of the damage to wildlife. It wasn't nice to be looking at it – birds, frogs, hedgehogs destroyed. The whole place was wiped out,' firefighter Breandán Ferriter told *The Kerryman*.[45]

Burning also creates hazardous local and regional air pollution and causes soil erosion that leads to flooding; it threatens local housing and costs millions annually in fire-fighting, as well as producing significant carbon emissions. This orgy of destruction is justified largely to facilitate heav-ily subsidized and loss-making sheep operations. Much of this burning is illegal, but prosecutions are rare and, even then, sanctions are derisory, with fines rarely exceeding €100. 'Prescribed fire can enable rapid and cost-effective treat-ment of unwanted vegetation,' according to Teagasc.[46] This is only true as long as you ignore the vast collateral damage this practice entails, and as long as the losses involved are not borne by those doing the burning. The Irish Wildlife Trust (IWT) has repeatedly called for all upland burning to be banned, citing the massive loss in biodiversity resulting from 'misguided public policy, unregulated turf extraction, over-grazing by sheep and uncontrolled burning'.[47] Fire is a traditional means of clearing vegetation for grazing. However, the highly degraded and overgrazed nature of many of our uplands makes them susceptible to burning. Rewilding, the IWT argues, would benefit wildlife and also be a buffer against fire and flood risk.

Another rarely discussed hazard associated with sheep are the toxic pesticides, cypermethrin and organophosphate compounds in sheep dips used to treat animals.[48] These chemicals pose a contamination risk to aquatic ecosystems in particular, and this effect is particularly acute in sensitive upland environments. The Department of Agriculture advises that used sheep dip be mixed with slurry or water and then spread on the land, the assumption being that its toxic effect is thus diluted, but this advice seems more about convenience than avoiding environmental contamination.

According to Teagasc, two in five sheep farmers had a family farm income below €5,000 in 2023. Average income for sheep farmers was €12,600, which is less than half the minimum wage.[49] Presumably in an attempt to address this, Budget 2025 saw a new payment of €13 per ewe introduced, at a cost to the taxpayer of €8 million.[50] This is in addition to the €12 per ewe already paid under the EU's Common Agricultural Policy. But it is perverse to subsidize activity that creates such huge financial and environmental costs for society. Rather than pay farmers to keep sheep, the state should pay them to de-stock and to act as stewards of rewilded upland landscapes.

With the removal of commercial sheep farming from many of our uplands and all our national parks, as well as aggressive measures to control deer populations, much of Ireland's uplands would, within a decade or two, begin to naturally regenerate with trees and, eventually, a self-restored forest canopy. Minimal state effort is required, other than fencing off and eliminating invasive herbivores from our mountains and uplands. An ambitious programme of native forest restoration for Ireland should see us aim for at least a quarter of our land to be forested by mid-century, bringing major climate and biodiversity gains.

*

The term 'national park' is used worldwide, and has specific meaning. According to the International Union for Conservation of Nature, national parks should 'protect large-scale ecological processes, along with the complement of species and ecosystems characteristic of the area'. While the concept of pristine nature may be utopian, surely our national parks should at least tilt the balance in favour of natural systems, while removing or at least significantly limiting negative human impacts.

In Ireland, it seems, we take a different view. Within forty-eight hours of the high-profile announcement of Páirc Náisiúnta na Mara, comprising a number of islands and coastal areas in Co. Kerry, minister Malcolm Noonan assured Kerry farmers that they would not face 'any additional burdens' as a result of the designation.[51] Conor Pass – which is part of the new national park – today resembles a moonscape of low scrub, with barely a tree anywhere in sight. This is the result of decades of overgrazing. Despite its new status as part of a state-owned national park, it appears that nothing substantive is going to change. Housing minister Darragh O'Brien, who was involved in the announcement of the new national park, said: 'Anything that took place yesterday will take place today and take place tomorrow as well. There's no additional restrictions in any way, shape or form.'

Ireland now has a total of seven national parks. Ecologist Pádraic Fogarty visited six of them on multiple occasions between 2013 and 2016 while researching his book *Whittled Away: Ireland's Vanishing Nature*. His account is harrowing: ecological wreckage, erosion, burning, invasive species, industrial timber plantations and disastrous overgrazing by deer, sheep and goats. All this, bear in mind, is occurring *inside* our so-called national parks.

While sheep farming on the hills has been carried out for centuries, things took a turn for the worse during the 1980s and 1990s, when EU payments to farmers were linked to the number of animals they kept. In Galway and Mayo alone, Fogarty notes, the sheep population doubled between 1980 and 1989, to over half a million, with disastrous consequences for the fragile ecology of the uplands.

Ireland long ago eradicated the apex predators such as wolves and lynx that would otherwise manage landscapes and prevent herbivores from overgrazing. 'Trophic cascade' is the phenomenon whereby entire ecosystems are disrupted and diminished by the removal of top predators. Growing awareness of its negative impacts has seen a concerted effort across many European countries to reintroduce apex predators. Grey wolves were virtually extinct in Europe but in recent decades their numbers have rebounded to an estimated 17,000.[52]

In September 2022, a wolf killed a thirty-year-old pony called Dolly in a paddock in Lower Saxony, Germany. A year later, the pony's owner, European Commission president Ursula von der Leyen, initiated an EU-wide review of the conservation status of wolves, claiming they had become a danger to humans as well as livestock.[53] While the future of top predators in Europe remains in the balance, in Ireland to even mention the possibility of reintroducing predators is considered extremist. Green Party leader Eamon Ryan mooted the idea in the Dáil in 2019 and was roundly ridiculed for daring to breathe the notion.[54] Early in 2024, Green Party biodiversity minister Pippa Hackett admitted that, were this to happen in Ireland, 'they'd all be shot'.[55] She noted that domestic dogs kill sheep regularly in Ireland, yet no one is proposing eradicating dogs.

To date, there has been no concerted national rewilding effort in Ireland, but localized efforts hint at the scale of what is possible. Eoghan Daltun bought a small, dilapidated farm on the rugged Beara peninsula in 2009 and set about fencing it off from herbivores. The results, as he recounts in his book *An Irish Atlantic Rainforest*, were spectacular. In a matter of years, the dull monotones of an overgrazed landscape quickly gave way to a riot of diversity. 'The visual effect was one of wildly splashing impressionist colour about the blank canvas of a previously barren woodland floor. Slowly at first, but at an ever-increasing velocity thereafter, life in all its vibrancy was coming back to the woods.'

Farmers, Daltun notes, are 'being continually pushed by official policy to destroy wildlife habitat, when exactly the opposite is what needs to be happening'.

Apart from helping restore ecosystems to good ecological health, rewilding programmes could also be a major tool in emissions reduction, while helping to mitigate the ever growing risk of extreme flooding.[56] Intact uplands, wetlands, marshes and boglands draw down millions of tonnes of atmospheric CO_2, as well as keeping billions of tonnes of soil-based CO_2 safely locked away. An Irish rewilding programme that placed a true value on emissions reduction, flood risk abatement and ecological restoration could be a

critical first step towards addressing our most acute environmental issues. This will only succeed with the co-operation
and participation of landowners, and selling its multiple benefits while allaying genuine concerns that farmers may have,
including of economic losses, remains a formidable challenge.

The prospect of rewilding, even at a micro-level, has
gained dramatic traction in Ireland in recent years. This must
begin with education and awareness. New organizations like
Farming for Nature are helping farmers who want to restore
nature on their lands. What's good for nature is also good
for farmers; the benefits of nature-friendly farming include
reduced risk of flooding and drought, more pollinators and
lower overall input costs, since healthy soils are naturally more
productive.[57] Many local authorities have reduced or eliminated chemical pesticides along roadside verges and in public
parks, while initiatives like 'No Mow May', which encourages the public to avoid mowing their lawns until June at the
earliest, are proving extremely popular.[58] It takes conscious
effort to overcome our cultural inclination to tidy up nature.
Neatly mown lawns, tightly trimmed hedgerows and smooth,
even surfaces eliminate the nooks and crannies, the unkempt
edges and the messy margins that nature clings to. Learning
to appreciate that such messiness is a sign of attention, not
neglect, is part of a journey towards ecological awareness that
many of us – urban and rural – are only just embarking upon.

It's probably too soon to say for sure what impact measures like these are having, but I did notice a welcome return
in early summer 2025 of some 'bug splat' on my car wind-
screen when venturing down the country.

*

If, when you think of an Irish bog, the image of red-headed
children and a donkey laden with hand-cut sods come to

mind, then you, like millions of people around the world, have seen the work of the twentieth-century photographer John Hinde. His iconic images, sometimes seen as kitsch, have framed a romantic notion in the public imagination of turf cutting as both timeless and benign. Were a latter-day Hinde to take his camera to an isolated raised bog, instead of a solitary donkey, he would be more likely to encounter a 200-horsepower industrial excavator ripping up the land. Impressions linger, however, and Ireland's ancient boglands have paid a heavy price for this collective false memory.

Ireland has around 1.4 million hectares of bogland, or around a fifth of the island – more in percentage terms than any other European country except Finland.[59] But the Ireland I grew up in did not value bogs. In school, to be called a 'bogman' or 'bogger' indicated you were backward or poor. Boglands were bad lands, and huge efforts went into draining and improving peaty soils to make them suitable for grazing or tillage. There was, until relatively recently, little or no awareness of the true value of peatlands and the major downsides of destroying them.

Ireland's peatlands lock away an estimated 1.2 billion tonnes of carbon. They are also home to around half our endangered bird species and around a quarter of our endangered plant species. A healthy bog typically stores ten times more carbon per hectare than any other system, including forest.[60] Peatland protection, according to the UN Environment Programme is 'among the most cost-effective options for mitigating climate change'.

As the Emergency ended, the Turf Development Act 1946 brought Bord na Móna into existence, with the semi-state body mandated to scale up peat extraction using mechanization. This continued until the end of 2020, when Bord na

Móna announced it had ended peat extraction, or 'harvesting', as it is often incorrectly described. It takes a healthy bog typically 1,000 years to add one metre in peat depth, so removing this is in no way 'renewable' in human timescales.

While peat may have had a role in helping secure Ireland's energy needs during wartime, the continuation of this almost to the present day was little short of scandalous.[61] Three peat-powered stations were commissioned in the midlands in the early 2000s and at their peak were burning around 3 million tonnes of peat annually. These were so inefficient that their operation depended on up to €100 million a year in subsidies via the Public Service Obligation.[62] This was an outrageous misuse of public funds to prop up what was little more than a politically engineered job-creation exercise in the midlands. Apart from being inherently inefficient, peat produces on average three times more emissions per megawatt of electricity generated than coal, which is itself an extraordinarily dirty fossil fuel.

Bord na Móna has rehabilitated around 20,000 hectares of bogland to date, or around a quarter of the area degraded by its industrial activities.[63] Its Peatland Climate Action Scheme aims to rehabilitate a further 33,000 hectares, a major undertaking that will take years to complete.

While Bord na Móna has been moving in the right direction, there remains a massive problem of illegal peat extraction from raised bogs located in special areas of conservation (SACs). Turf cutting in these areas was formally banned in 2011, and over €65 million has been paid in compensation to 3,435 turf cutters – an average of almost €19,000 each.[64]

According to information obtained by the Irish Wildlife Trust, some people who were compensated continue to cut turf with impunity.[65] Some 330 plots on SAC bogs were cut in 2022.[66]

In terms both of their biodiversity and importance for climate change, it is no exaggeration to describe Ireland's bogs as our rainforests. After decades of deliberate destruction as well as neglect, by state and private companies and by individuals, there is at last emerging an understanding of the true value of these unique habitats, among the very last of their kind to survive in Europe.

The European Commission is to refer Ireland to the European Court of Justice for our failure to apply the Habitats Directive in protecting designated raised and blanket bogs from extraction. 'These areas are biodiversity hotspots playing host to important insect and bird species,' the Commission wrote. 'They are categorized as "priority" habitats under the directive due to their unique qualities. Peat bogs are also vital carbon sinks when healthy, while Ireland's degraded peatlands emit 21.5 million tonnes of CO_2 equivalent per year.'[67] Despite our clear legal obligation to protect these special areas of conservation, rural TDs strongly support continued turf cutting, and politicians at national level are running scared of opposing these interest groups head-on.

All our boglands really need from us is to leave them alone, recognizing their dual role in climate abatement and biodiversity recovery. The Irish state must face down those who, for personal gain, continue to flout the law in draining and destroying these sensitive protected habitats.

*

Nobody can say for certain what the future will be like, but there are many reasons to suspect that it will be more like the deep past than the present, as our age of hyper-abundance and overconsumption careers towards an abrupt end. This point was brought home forcefully during the launch of the 2024 report of the *Lancet* Countdown on Health and Climate Change.[68] One of the co-authors, Karyn Morrissey of the University of Galway, warned that children born in Ireland today may well face hunger within their lifetime. Climate-fuelled extreme weather events are taking an ever more severe toll on food systems. 'I'm really worried about col-lapses in food supply because our farmers aren't able to grow crops, and not just our farmers but across Europe,' Morris-sey said. The future Morrissey and her peers on the *Lancet* Countdown project are warning of will involve, as she put it, 'the eradication of current living standards'.[69]

This is what is coming down the line. We still have time to hunker down and brace for the approaching climatic storm, but only just, and only if we act now and act decisively. Our current system of producing crops to feed livestock largely for export, while importing four-fifths of everything we eat, contains the seeds of catastrophe.

The model of agroecological farming, which applies sci-entific principles to work with nature and protect ecosystems while achieving domestic food security and breaking our dependence on imported inputs, offers Ireland a way for-ward in the tumultuous decades ahead. It is being pioneered by groups like Talamh Beo and Farming for Nature, but so far they are ploughing a lonely furrow, with minimal fund-ing and only lip-service from politicians by way of support. Real strides have been made in expanding the share of Irish land being farmed organically, but much of this transition has been among livestock farmers, while Ireland's tillage and horticulture sectors remain in long-term decline.

As we enter a global Long Emergency, what's now needed is the establishment of a Department of Food Security that is free to set agricultural policy and direct resources to secure Ireland's domestic food supply in a world of deepening geopolitical uncertainty, deglobalization and climate crisis. There are fewer people involved in agriculture today than at any previous point in our history, with barely one in twenty-five of the total labour force working the land. Farms have grown bigger, more specialized and more mechanized, leading to a huge decline in employment even as output has increased. The average age of Irish farm owners is almost sixty; 40 per cent are aged over sixty-five.[70] This age profile, which has significantly worsened in the past decade, represents an opportunity: if even a small fraction of the land currently dedicated to dairy and beef production were transferred to horticulture and agroecology as the landholders retired, this would provide more than enough land to develop a scaled-down version of the Dutch model of intensive horticulture.

A new vision for food and land in Ireland should entail a sharp increase in horticulture, tillage and agroecological systems focused on meeting local and national food needs first, with the surplus being exported. These types of farming are much more labour-intensive than beef-cattle rearing or dairying, which means an influx of younger farmers and farm workers is needed in the coming years. This would have the effect of regenerating rural Ireland and reversing the long-term flight from the land.

The virtual elimination of chemical fertilizers would reduce the current artificially high carrying capacity of our grasslands back to the levels of the 1960s. This in turn would mean a sharp reduction in cattle numbers as well as the almost complete elimination of the current annual need to import up to 5 million tonnes of animal feed, almost 40 per cent of which is consumed by dairy cattle. Much of this feed comes from South America.[71] According to Teagasc, South American maize creates twenty times the level of emissions per hectare as domestic tillage, yet Ireland is now importing millions of tonnes of maize and soya from areas suffering deforestation. In a rapidly warming, food-stressed and politically unstable world, it is highly unlikely that reliable flows of cheap imported fodder or millions of tonnes of affordable chemical fertilizers to maintain our oversized livestock herd can be sustained into the future.

Beef production, too, is staggeringly inefficient. According to the Global Food Policy Report, 'Only 1 per cent of gross cattle-feed energy and 4 per cent of ingested protein are converted to human-edible calories and protein.'[72] Apart from being woefully inefficient at food production, beef farming in Ireland is completely dependent on taxpayer subsidies, which make up 139 per cent of cattle rearing farm incomes.[73] And, of course, these subsidies are perverse, inasmuch as they perpetuate an industry that creates such vast environmental costs for our country.

The science says that if we wish to avoid climate catastrophe, we have no choice but to pivot decisively towards mainly plant-based diets. 'Even with ambitious action to reduce the emissions intensity of livestock production, it is unlikely that global temperature rises can be kept below two degrees Celsius in the absence of a radical shift in meat and dairy consumption,' according to a Chatham House study.[74] It remains to be seen to what extent humanity – or the part of humanity that lives in the rich world and consumes a grossly disproportionate share of the world's meat and dairy – will make this necessary transition. But any intelligent and responsible strategy for the future of Irish agriculture has to be rooted in an acceptance that it is senseless to continue with a food production model dominated by meat and dairy. Even those who do not accept the moral and economic reasons for this will soon have to bow to the reality that the Irish agricultural model simply cannot be sustained.

What is needed is a dramatic reorientation of agricultural subsidies, away from these inefficient and destructive sectors and towards sustainable forms of food production. The transition away from cattle and sheep and towards tillage and horticulture would free up millions of hectares of farmland for a variety of other uses, including rewilding.

A future for rural Ireland could be one of thriving new communities engaged mainly in plant-based food production, much of it under glass and in polytunnels. There would still be a role for livestock in this mixed farming model, as there always has been, but it would no longer be dominant.

The future is what we make it. We can lumber on with the same monocultural livestock-dominated systems that have wreaked havoc on our biodiversity, degraded our waterways and ramped up climate-damaging emissions. Or we can set a new course for a future Ireland: globally responsible, food-secure, and with a repopulated countryside. We can save ourselves years of needless hardship and suffering if we accept reality now and grasp the challenge and work with the farming community to begin a great transition to a new food and land system in Ireland.

9. Adapt or Die

What does the future hold for Ireland's climate? The TRANSLATE project, led by Met Éireann, has produced the first standardized national climate projections for Ireland.[1] This involves supercomputer modelling of how our climate would react to increases in global temperatures ranging from 1.5 to 4°C above pre-industrial levels. The modelling paints a picture of an increasingly fractious climate system, with much wetter winters, more frequent droughts, heatwaves and downpours, and more powerful storms.

As a mid-latitude country, Ireland is warming roughly in line with global averages. Over the last thirty years, Ireland has become markedly warmer and wetter: average air temperature rose by 0.7°C and there was an overall increase in rainfall of around 7 per cent.[2] These are astonishingly rapid climatic shifts. Research led by Conor Murphy of Maynooth University found that the decade from 2006 to 2015 was the wettest in over 300 years in Ireland.[3] In that period, Ireland's average annual rainfall was almost 1,990mm, nearly double the long-term average.

Since the early 2000s, Ireland has experienced an extreme weather event, on average, every six to eight months. This represents a four- to five-fold increase in frequency versus Irish weather several decades ago, according to climatologist Kieran Hickey.[4] In June 2023, a category-IV marine heatwave occurred in the north Atlantic off the coast of Ireland, with some areas recording a category-V heatwave, conditions described by scientists as 'beyond extreme'.[5] Warmer waters provide more energy for storms and can also contribute to extreme rainfall events and cyclones.

While extreme heatwaves and drought conditions threaten many parts of the world, for Ireland, extreme precipitation and flooding events are our greatest near-term climate hazards. Ireland is, as John Sweeney has written, 'at the mercy of what happens around us in the Atlantic Ocean'.[6]

Since the beginning of the twenty-first century, Ireland has endured a series of significant flooding events on an almost annual basis.[7] The rest of this century will be a rolling battle to hold back water, be it from rising sea levels, violent storm surges or increasingly intense downpours. In many cases, this will involve a staged retreat from areas that are impossible or simply too expensive to defend.

The Office of Public Works (OPW) has principal responsibility for design and delivery of flood protection and relief projects. It operates under 1945 legislation that was primarily intended to improve land for agriculture via drainage schemes. Back in the mid-twentieth century, the underlying assumption was that the climate would remain broadly stable. That is demonstrably no longer the case, and the OPW acknowledges this in noting that all new flood relief schemes are designed 'to take account of climate change'.[8]

The unavoidable fact facing all such schemes is that water has to go somewhere. If we prevent rivers from spreading into their natural floodplains, and if land management practices degrade the ability of our uplands to retain rainfall, torrents of water will instead be channelled downstream, with dangerous consequences.

In some instances, flood protection measures can actually

make flooding worse: for instance, man-made dykes, once breached, can actually prevent flood waters from returning to the river. John Sweeney believes we should be paying farmers to allow periodic flooding of their low-lying fields, in place of the old approach of arterial drainage.

Valuable lessons can be learned from the small market town of Pickering in North Yorkshire, once a notorious flooding blackspot. Between 1999 and 2007, the town was flooded four times.[9] Pickering was refused a proposed £20-million flood defence scheme on the grounds of the cost being too high to justify protecting a limited number of people. In any event, the large concrete works in the proposed scheme would have been an eyesore in a town that depends on tourism income. Instead, a panel of academics, environmentalists and experts from the UK Environmental Agency and Forestry Commission were assembled to study the root cause of the repeated flooding. They came up with a number of nature-based solutions, including hand-built leaky dams made of logs and branches, as well as planting woodlands and halting the burning of uplands. The total cost was £2 million, or one-tenth of the cost of the proposed concrete wall for the town centre. Since these solutions were implemented, Pickering has been flood-free. The success of this intervention has since been replicated in other communities in Britain.

'Pluvial' flooding – directly from intense rainfall – is becoming an ever greater threat in urban areas. Cities are losing permeable surfaces as more land is covered in concrete and asphalt. A UCD study found that in just a two-year period, Dublin city lost around 5 per cent of its permeable surfaces.[10] A major cause of this loss is the paving of gardens for car parking or to reduce maintenance. While Dublin

City Council is aware of the issue, there are no regulations in place to compel homeowners to retain permeable surfaces on their property.

Dublin's drainage infrastructure, built in the Victorian era, was not designed to cope with today's more intense rainfall levels. Retrofitting the entire system would be nigh-on impossible, so the only realistic option within built-up areas is to protect existing permeable surfaces and, where possible, expand them.

Ironically, given that Ireland is wet and getting wetter, our supplies of drinking water are under increasing pressure. As climate change intensifies, the heavily populated eastern region is becoming dryer while the west is getting wetter. In early 2025, Uisce Éireann chairman Jerry Grant said that the state's water and sewerage systems 'are in a desperate state' because of a lack of investment in infrastructure.[11] To compound matters, Ireland, uniquely among developed countries, does not charge the public for water. Up to €6 billion may need to be spent to transfer water from the Shannon to the greater Dublin area, where droughts and water shortages are expected to become a critical risk.

*

When it comes to man's mastery over nature, Holland stands apart. For centuries, it has reclaimed land from the sea and constructed a vast network of dykes to protect its settlements and farmland. Around a third of the country is below sea level, and almost two-thirds of Dutch homes are in areas at risk of flooding.

The country's vulnerability to the elements was cruelly exposed in January 1953, when a severe storm tide led to over 1,800 fatalities, with nearly 50,000 buildings damaged, 140,000 hectares of land inundated and tens of thousands of livestock drowned.[12] Since then, its defences have been greatly strengthened. But rising sea levels will expose the limitations of 'hard' engineering solutions. Recognizing this, the Dutch ministry for infrastructure and water management has initiated a programme called 'Room for the River' at thirty locations around the country.[13] The aim is to restore rivers' natural floodplains to help relieve pressure on the dykes. One-fifth of the project's total budget of around €2.4 billion has been spent on buying out and relocating over 200 households. Some dykes have been removed, and in one project around twenty-seven square kilometres of land has been given back to the river entirely. The effect of this initiative has been to greatly reduce flooding risk for more than 60,000 people in the region around the nearby town of Werkendam. Much farmland remains in the area, but it is now designed for occasional flooding. These interventions have had the welcome side effect of creating new havens for wildlife: swans, geese, ospreys and sea eagles have recolonized the newly created wetlands.

A bigger challenge was faced in Nijmegen, a city similar in size to Cork. Twice during the 1990s, around a quarter of a million people were evacuated from Nijmegen and its surrounding areas amidst fears the River Waal would burst its banks and inundate the city. The solution was to clear an area of around 250 hectares to create a new floodplain. At the insistence of the local population, this land has been turned into a civic amenity area, with cycle paths, pedestrian walkways and green spaces. The first big test of the project came in 2018, when nearby German towns protected by dykes were flooded by the Waal but Nijmegen remained dry.[14]

A local community group in Skibbereen, Co. Cork, proposed a similar approach: the development of an environmental

park to serve as a flood storage basin. The idea was rejected by
the local council, which instead opted for a €14-million hard-
engineering approach, as it preferred to develop the area in
question as a car park.[15]

The Dublin suburb of Clontarf was badly affected by
a major flooding event in 2002. In response, Dublin City
Council developed a plan to protect a vulnerable three-
kilometre stretch of coastline with a combination of flood
wall and earthen embankment, which was put in place in
2017. The following year, after a protracted dispute with
local residents, who claimed it partially obstructed sea
views, the council spent half a million euros lowering a 500-
metre-long section of the sea wall by thirty centimetres.
The council's chief executive warned that the wall would
need to be raised again 'at some future date'.[16] In 2020, the
council developed new plans for enhanced flood defence
works at Clontarf, but it is unlikely that permanent protection will be put in place before 2027.[17] Meanwhile, every
year that passes rolls the dice on another flooding disaster
in Clontarf.

'Most people in Ireland hopelessly misunderstand risk,'
climatologist Peter Thorne told me. 'You see this in the
number of people who buy properties, and the number
of local authorities who provide planning permission for
properties, in flood-prone areas.' The controversy over the
Clontarf sea wall is, he added, 'the poster child of this;
people do not want to have their views spoiled, and believe
that somehow they will not get flooded'. Thorne is bewil-
dered at the sluggish pace of climate adaptation in Ireland.
While we have to play our part in global efforts at climate-
change mitigation, on adaptation the responsibility is ours
and ours alone. 'If we fail on adaptation, it's a failure of the

government to protect its citizens, its businesses, hospitals, schools and everything else.'

The standard refrain from politicians across the spectrum is that climate action should be about incentives rather than penalties. Thorne argues that this is a major reason why climate action in Ireland continues to be inadequate. 'We are still dealing almost exclusively in proverbial carrots and not dealing in sticks,' he told me. Off the record, politicians will tell you they know exactly what steps need to be taken to get serious about meeting tough emissions targets, but none know how to get re-elected afterwards.

Semi-state bodies including Dublin Port, the Dublin Airport Authority and Irish Rail have statutory reporting obligations around mitigation but not adaptation, Peter Thorne notes. In the case of Irish Rail, he says, 'what you find is they buy shiny new trains that are lower emissions, and they fail to take into account that the railway line from Dublin to Rosslare has several points that are about to fall into the sea'. Along some parts of the east coast rail line, twenty to thirty metres of coastline has been lost in the past decade. The East Coast Railway Infrastructure Protection Projects were established to address this fast-emerging vulnerability.[18] The aim is to increase resilience and protection from climate-driven erosion in five zones from Merrion in south Dublin to Wicklow harbour.[19]

It is unclear whether critical rail infrastructure so close to a rapidly eroding coastline can realistically be protected in the longer term, or whether the focus should instead be on establishing an alternative inland route. Irish Rail chief executive Jim Meade admitted in 2023 that the Dublin–Rosslare railway line is now 'under real threat'. As the *Business Post* reported, an earlier proposal to move the line inland, including a stretch along the cliff face between Greystones and Bray, was rejected by Irish Rail on the basis of cost and because it would have involved infringing on private property.[20]

There is at least some rationale in fighting to protect essential infrastructure such as rail lines. On the other hand, it makes no sense whatever trying to protect farmland along the coast. The cost of such protection, even were it feasible, would be ten to a hundred times the value of the land, according to John Sweeney. Ad hoc efforts at coastal protection by landowners can have highly negative consequences, with wave energy being redirected elsewhere.

In late 2024 the Climate Change Advisory Council published a working paper on the economic costs of climate impacts and adaptation.[21] Unless strong adaptation measures are implemented, it estimated that climate impacts in coastal areas would cost around €2 billion a year by mid-century, with coastal flooding and inundation the biggest risk. A more surprising projection of the study was that even apparently modest temperature increases would lead to a significant uptick in emergency hospital admissions in Ireland. Its overall conclusion is that it makes overwhelming economic sense to invest heavily in adaptation policies now, as their cost is only a fraction of the losses we will otherwise be exposed to. Buy now, in other words, or pay later.

In the event of one metre of sea-level rise, around 350 square kilometres of coastal land is 'vulnerable to flooding', increasing to 600 square kilometres at three metres of sea-level rise.[22] While these may not sound like huge areas, they include much of our high-value coastal infrastructure and settlements. There is vast uncertainty as to how sea-level rise will progress this century, but even in a best-case scenario, long-term sea-level rise over the coming centuries is already locked in. The last time the planet was this warm and global CO_2 levels were similar to today was during the mid-Pliocene warm period around 3 million years ago.[23] Global sea levels at that time were twenty metres higher on average than today.

*

In late 2024, the Central Bank reported that up to 100,000 Irish homes – around one in twenty – were struggling to secure flood insurance.[24] It estimated that the annual cost of flood damage to buildings now exceeds €100 million, with the bulk of that affecting uninsured buildings.

The Central Bank report highlighted the extensive flooding associated with Storm Desmond, which hit Ireland's west coast in December 2015, with losses of over €200 million. The same system caused far more severe damage in northern England, where it stalled for three days. This was, the report noted, a 'near miss' for Ireland. Had Storm Desmond tracked over Ireland's much more densely populated east coast, property damage from inland flooding was modelled to have exceeded €1.1 billion. Storms of this magnitude are now likely to occur every eleven years, and will put 'significant financial strain on both the insurance industry and the State'. A Central Bank survey found that only one in five insurers fully account for climate risk, while fewer than half have adequately stress-tested their financial exposure to climate risks.

In May 2024, EU commissioner Mairead McGuinness noted that, in recent decades, only a quarter of the total losses arising from extreme weather events in Europe were covered by insurance.[25] As climate change intensifies, insurers and reinsurers were, she warned, likely to retreat from providing cover, leaving governments and taxpayers exposed to covering the gap. 'Investing in climate action in a timely way will reduce this huge risk to taxpayers,' McGuinness said.

A study by the Irish Fiscal Advisory Council warned that extreme weather events could 'threaten the financial stability of insurance firms', forcing them to sell assets to meet claims. An actuarial report in 2023 found serious flaws in the economic models used to assess financial risk around climate change, including for insurers, as these failed to account either for compound risks or for climate tipping points.[26] Eric Anderson, president of insurance giant Aon, commented that 'just as the US economy was over-exposed to mortgage risk in 2008, the economy today is over-exposed to climate risk'.[27]

Climate risk could well be the domino that triggers the next global financial crash. Without a functioning insurance market, the property sector cannot operate, as banks will not issue mortgages on uninsured or uninsurable properties.

In March 2025, a senior insurance-industry figure summarized the threat posed by climate change in a LinkedIn post titled 'Climate, Risk, Insurance: The Future of Capitalism'. Günther Thallinger, a board member of Allianz SE, wrote: 'We are fast approaching temperature levels – 1.5C, 2C, 3C – where insurers will no longer be able to offer coverage for many of these risks. The math breaks down: the premiums required exceed what people or companies can pay. This is already happening. Entire regions are becoming uninsurable.' Thallinger sketches a cascade of consequences for the wider financial system, from a 'climate-induced credit crunch' and 'market failure' to a point – at global temperatures 3°C above pre-industrial levels – where 'capitalism as we know it ceases to be viable'. He concludes: 'There is only one path forward: prevent any further increase in atmospheric energy levels.

That means keeping emissions out of the atmosphere. That means burning less carbon or capturing it at the point of combustion. These are the only two levers. Everything else is delay or distraction.'

*

'Prediction is very difficult, especially if it's about the future,' quipped the Danish physicist Niels Bohr. What has been striking about the predictions made by climate scientists is that they have, to an overwhelming degree, been borne out by reality – except, in many cases, with even greater speed and intensity than the models projected. We now face a range of possible futures on a spectrum running from extremely challenging to outright catastrophic.

While the consensus of evidence points to our climate growing both warmer and wetter in the coming decades, there is also another possible pathway for Ireland.

The main system of currents in the Atlantic Ocean is known as the Atlantic Meridional Overturning Circulation, or AMOC. This vast system of ocean currents is the reason why Ireland, Britain and much of north-western Europe, despite being at quite high latitudes, enjoy a relatively mild climate. At 53° north, Ireland is on the same parallel as Labrador and Newfoundland, where mean winter temperatures remain well below 0°C from December through to March. Ireland's coldest months, in contrast, are January and February, with mean temperatures of 4–5 °C; relative to Newfoundland, this is positively balmy.

The main reason for the difference between Newfoundland and Ireland is the AMOC, which transfers around one peta-watt of heat energy from the tropics deep into the northern hemisphere. This stupendously large energy transfer is equivalent to the total output of a million nuclear power stations.

While there have been repeated warnings that the AMOC is weakening, having lost some 15 per cent of its power in recent decades, it was less clear whether it would continue to weaken, or suddenly shut down. Research published in early 2024 reported on the findings of a massive climate modelling project based on data from direct observations that took a Dutch supercomputer facility six months to run. It concluded that 'the present-day AMOC is on route to tipping'.[28] This study identified what are called early warning indicators that a full AMOC shutdown may be approaching. Such a tipping event has not happened for more than 10,000 years, and it would have catastrophic and far-reaching consequences, including triggering a one-metre sea-level rise in the north Atlantic. Drastic cooling in Europe would devastate our food production systems. 'What surprised us was the rate at which tipping occurs. It will be devastating,' according to René van Westen of Utrecht University, the study's lead author.[29]

On a recent visit to Dublin, oceanographer Stefan Rahmstorf of Potsdam University in Germany told me: 'What's important to bear in mind is that once it shuts down, the AMOC is, in human timescales, gone forever.' His message to policymakers and politicians was blunt: 'They simply have to start putting climate at the top of their agendas; they cannot allow it to be pushed off the agenda by other short-term priorities. This is an extremely urgent situation.'[30]

Rahmstorf was one of more than forty international experts who co-authored an open letter to the Nordic Council of Ministers in late 2024, warning that the risk of major ocean circulation change in the Atlantic has been greatly underestimated, as have the 'devastating and irreversible impacts' of such an eventuality, including on the viability of agriculture in north-western Europe, obviously including Ireland.[31]

An AMOC shutdown has the potential to throw an enormous spanner in any efforts already underway to cope with a warming climate. Measures taken to adapt to a warmer, wetter near-term future in Ireland and beyond could turn out to be completely unsuited to a much colder, dryer post-AMOC regime. Some scientists admit to being reluctant to discuss an AMOC shutdown, for fear that it could actually stymie current climate action. 'If we give politicians what they call the two-handed economics answer – on the one hand it might be this, and the other hand it might be that – it gives them the clear goal of doing nothing, and that's what you don't want to happen,' John Sweeney told me. While pointing out that the IPCC's latest report expresses 'medium confidence' that an abrupt AMOC collapse will not occur before 2100, Sweeney accepts that this assessment may itself already be out of date, as a slew of recent research paints an altogether more alarming picture.

In the immediate future, Sweeney's major concern is focused around climate-induced changes to the jet stream, the narrow high-altitude and fast-moving currents of air that in the northern hemisphere mainly flow from west to east, driven by the Earth's rotation. Rapid heating in the polar regions is thought to be making the jet stream 'wobble', its pattern becoming increasingly erratic. This in turn is fuelling stalled weather patterns that allow the development of intense heatwaves, as well as slow-moving storm systems that result in devastating flooding events.[32]

In 2016, the pioneering US climatologist James Hansen and colleagues published a research paper warning of drastic consequences should the ocean conveyor system fail.[33]

The sudden increase in the temperature gradient between the tropics and the northern hemisphere would, according to Hansen, 'drive superstorms stronger than any in modern times. All hell would break loose in the North Atlantic,' he added. Hansen's research team identified a previous period of rapid global warming around 120,000 years ago, towards the end of the Eemian period. At that time, a collapse of polar ice, swiftly rising sea levels and a shutdown of the AMOC led to superstorms of epic dimensions in the region. Scientists have hypothesized that evidence for this event exists in rocks, weighing around 1,000 tonnes each, that are located atop a cliff twenty metres above the Atlantic in the Bahamas. These giant boulders may have been scooped up from the sea floor and tossed inland during a period of unimaginably powerful megastorms far beyond anything modern humans have ever endured.

*

As the world heats up, species are migrating away from the tropics and towards cooler climes. But for many species, the climate is changing far more quickly than they are capable of adapting to.

Humans are, of course, among the species being uprooted. In 2022, a record 32.6 million people were forcibly displaced as a result of floods, storms, wildfires and droughts. The Institute for Economics & Peace (IEP) calculates that by 2050, upwards of a billion people will be climate refugees, driven from their homelands by environmental breakdown and the civil unrest and armed conflict it will trigger.[34] In the nearer term, the IEP projects that by 2040 around 5.4 billion people, over half the world's population, will live in regions experiencing high or extreme water stress. As rivers often flow through multiple countries, the scope for geopolitical tensions and warfare over access to water is self-evident.

I had the opportunity to see these impacts at first hand when, in early 2020, I travelled to Zambia to report on severe water and electricity shortages the country was facing as a result of persistent drought.[35] In the three preceding years, water levels in Lake Kariba, the world's largest artificial lake, had dropped by six metres. This severely limited hydro-electric production by the Kariba dam, which provides around half of the total electrical supply for Zambia and Zimbabwe. Years of severe drought were followed by torrential downpours that destroyed crops and wrecked infrastructure that Zambia can ill afford to repair or replace. Energy shortages and crop failure have led many rural Zambians to cut down trees to sell as charcoal. As a result, over a quarter of a million hectares of forest are being cleared a year. 'We don't like to cut the trees but we have no choice,' Diana Moono, a roadside charcoal seller, told me. The outrageous injustice of climate change can be understood when you consider that Zambia's population of around 20 million people produces, in total, around 9 million tonnes of greenhouse gases annually, or less than one-sixth of the climate-destroying carbon pollution that Ireland's 5.3 million people account for.[36]

As politicians in the global north plan ever more draconian measures to prevent desperate migrants from reaching our shores, many of those same politicians, without apparent self-awareness, also oppose strong climate action. The refusal of rich countries like Ireland to act decisively on decarbonization and emissions reductions in line with the science may come to be seen as the ultimate own-goal, as climate-driven weather disasters and mass migration cause the international political and economic order to fray. Many of these risks were touched on in a report by the Irish Defence Forces in early 2025, which warned of widespread disruption to critical national infrastructure. In the report, Commandant Paul O'Callaghan warned that climate change 'has become a tool for far-right extremists to amplify xenophobia, anti-globalism, and anti-immigrant rhetoric, capitalizing on fears of migration, economic insecurity and resource scarcity'. What O'Callaghan described as a dangerous mix of ideologies 'requires close attention as both climate change and far-right extremism continue to shape global politics'.[37]

At least for now, Ireland is not subject to the extreme droughts and heatwaves racking much of the European continent. This gives us precious time to adjust to the coming changes. Whether we use this opportunity to weatherize our society as far as possible and to engage in a rapid programme to achieve energy independence and domestic food security, or whether we squander it and continue to stumble forward blindly into an increasingly uncertain and dangerous future, remains to be seen; the evidence thus far suggests we remain firmly on the latter path, essentially trading present comfort for future safety.

What would have to happen for the climate crisis to become our single dominant social, political and media issue? Nobody knows for sure, but it's likely to be pretty brutal. Since it hasn't happened yet, it's left to the creatives to try to imagine it. The opening chapter of Kim Stanley Robinson's 2020 novel *The Ministry of the Future* harrowingly chronicles a deadly heatwave in India in the near future that kills millions, an event that becomes the global catalyst for profound political, economic and social upheaval. Just such a catastrophe was narrowly avoided in 2022, and again in April 2025, when temperatures in densely populated regions of India and

Pakistan edged close to levels that could trigger a mass fatality event as persistent power outages reduced public access to air conditioning.

'We are on the brink of an irreversible climate disaster. This is a global emergency beyond any doubt. Much of the very fabric of life on Earth is imperilled. We find ourselves amid an abrupt climate upheaval, a dire situation never before encountered in the annals of human existence.' This was the sobering conclusion to a state of the climate report, endorsed by more than 15,000 practising climate scientists and published in late 2024.[38]

Despite the warnings, humanity remains in thrall to what philosopher Roman Krznaric – author of books including *History for Tomorrow* and *The Good Ancestor* – described as 'pathological short-termism'. We treat the future, he argued, 'like a distant colonial outpost devoid of people where we dump our ecological degradation, risk, nuclear waste and public debt'.[39] Perhaps what Ireland needs is a real-life Ministry for the Future? Sweden established just such a ministry in 2014, while the following year, our nearest neighbour, Wales, appointed a Future Generations Commissioner to help embed longer-term thinking into the political process.

The closest thing Ireland has managed to date has been the Citizens' Assemblies, our hugely successful experiment in deliberative democracy. They revealed, on issues as diverse as climate change and abortion, an Irish public more progressive, kinder and more empathetic than most people had imagined.

The eyes of future generations are on this generation, silently urging us to act bravely and quickly on the climate emergency, to save them from an immiserated future in a collapsed biosphere. Will we rise to the moment and earn the right to be regarded as 'good ancestors'?[40]

Acknowledgements

Researching and writing this book was a tougher challenge than I had expected, but also a great privilege. To those I interviewed or sought advice from over the past year and more, including many off-the-record conversations, your generosity with your time and insights is appreciated. I am especially indebted to Paul Price, an outstanding climate researcher, for his expertise and counsel around the chapters dealing with agriculture and food security, and to John Sweeney and Hannah Daly, who were generous with their time and advice.

This work would not be possible without my being able to draw on the reporting around climate and related issues in Ireland by a small cohort of specialist journalists. These include Lauren Boland, Philip Boucher-Hayes, Shauna Corr, Pádraig Hoare, George Lee, Ella Mc Sweeney, Daniel Murray, Caroline O'Doherty, Kevin O'Sullivan and Niall Sargent. To that list I would add the pioneering environmental broadcaster Duncan Stewart, the campaigning ecologist Pádraic Fogarty and, for his humour and wisdom in dark times, psychologist John Sharry. Among the many important environmental writers from beyond Ireland, I should single out George Monbiot, an inspiration to me and so many others.

My commissioning editor at Penguin, Brendan Barrington, had the unenviable task of trying to get me to stick to the point and to the brief. Without his critical skills this book might have been a great deal longer – and a good deal weaker.

In a stint on An Taisce's Climate Committee, I came to appreciate the vast breadth of pro bono work being undertaken in the wider public interest by environmental NGOs, often in the teeth of abuse and harassment. Those taking on this important work, either as volunteers or staff, deserve our gratitude.

To my business partner of more than thirty years, Geraldine Meagan, my thanks for her steadfast friendship and good-humoured tolerance of my extended absences from work.

My siblings Mary, Nuala, Ursula, Olive, Declan and Alan are a source of constant support and friendship, as is our wider family network. My mother, Ann, who died in 2022, always championed me in my writing and is sorely missed.

I am grateful to the late Cristóir Gallagher, my secondary-school English teacher, whose encouragement led me to consider a career in journalism.

I owe everything to my wife Jane, a talented and highly accomplished artist, the beating heart of our household and devoted mum to our daughters, two wonderful young people whose future is what keeps me in the fight. Jane has stood by me every step of the way on what has been an often rocky path through environmental journalism.

While this book is the fruit of much collaboration, inspiration and input from many of the people mentioned here, any errors or oversights are, of course, mine alone.

Notes

Chapter 1: Goldilocks Is Dying

1 Rebecca Lindsey, 'Climate Change: Atmospheric Carbon Dioxide', National Oceanic and Atmospheric Administration, 9 April 2024.

2 'The Atmosphere: Getting a Handle on Carbon Dioxide', NASA, 9 October 2019.

3 Xiaochun Zhang and Ken Caldeira, 'Time Scales and Ratios of Climate Forcing Due to Thermal Versus Carbon Dioxide Emissions from Fossil Fuels', *Geophysical Research Letters* 42(11), 2015: 4548–55.

4 'Atmospheric Greenhouse Gas Concentrations', European Environment Agency, 29 January 2025.

5 'WMO Confirms 2024 as Warmest Year on Record at About 1.55°C Above Pre-Industrial Level', World Meteorological Organization, 10 January 2025.

6 'Graphic: The Relentless Rise of Carbon Dioxide', NASA, 29 August, 2013.

7 Nicola Davis, 'The Anthropocene Epoch: Have We Entered a New Phase of Planetary History?', *Guardian*, 30 May 2019.

8 'Climate Change 2007: Synthesis Report. Contribution of Working Groups I, II and III to the Fourth Assessment Report of the Intergovernmental Panel on Climate Change', IPCC, 2007.

9 'Debunking Misinformation About Stolen Climate Emails in the "Climategate" Manufactured Controversy', Union of Concerned Scientists, 8 December 2009.

10 'Global Cooling', Editorial, *Irish Times*, 6 January 2010.

11 Timothy Morton, *Hyperobjects: Philosophy and Ecology After the End of the World* (University of Minnesota Press, 2013).

12 David Wallace-Wells, 'The Uninhabitable Earth', *New York*, 10 July 2017.

13 Zeke Hausfather, 'Analysis: Global CO_2 Emissions Will Reach New High in 2024 Despite Slower Growth', CarbonBrief.org, 13 November 2024.

14 'Ireland Is Projected to Exceed Its National and EU Climate Targets', Environmental Protection Agency, 27 May 2024.

15 Ibid.

16 'EPA Projections Show Ireland Off Track for 2030 Climate Targets', Environmental Protection Agency, 1 May 2024.

17 'Annual Review and Outlook for Agriculture, Food and the Marine 2024', Department of Agriculture, Food and the Marine, December 2024.

18 'Structural & Cohesion Policies – Agriculture & Rural Development', European Parliament Policy Department, January 2017.

19 Nick King and Aled Jones, 'An Analysis of the Potential for the Formation of "Nodes of Persisting Complexity"', *Sustainability* 13(15), 2021: 8161.

Chapter 2: The Lie of the Land

1 'Labour Market Economy Ireland and the EU at 50', Central Statistics Office, 17 October 2023.

2 James Humphreys, 'Nutrient Issues on Irish Farms and Solutions to Lower Losses', *International Journal of Dairy Technology* 61(1), 2008: 36–42.

3 Seamus Sheehy, 'A Long, Long History of Agricultural Reform', *Irish Times*, 22 February 1999.

4 'Frequently Asked Questions: End of Milk Quotas', European Commission, 26 March 2015.

5 'Food Harvest 2020: A Vision for Irish Agri-Food and Fisheries', Department of Agriculture, Fisheries and Food, 2010.

6 'Consultation on Environmental Analysis of Food Harvest 2020: Response of the Environmental Protection Agency', 2012.

7 Trevor Donnellan, Patrick Gillespie, and Kevin Hanrahan, 'Impact of Greenhouse Gas Abatement Targets on Agricultural Activity', 83rd Annual Conference of the Agricultural Economics Society, 30 March 2009.

8 'New Thinking for Agricultural Emissions Policies', *The Cattle Site*, 11 June 2012.

9 Harry McGee, 'Simon Coveney Says EU Has "Got It Wrong" on Carbon Emissions Targets for Agriculture', *Irish Times*, 2 September 2013.

10 Alison Healy, 'Simon Coveney Hails "Exciting" End to Milk Quotas', *Irish Times*, 20 November 2014.

11 'Finance Ireland to Increase Borrowing Limits for Dairy Farmers on Innovative Milkflex Product by 66 per cent to €500,000', TirlánFarmLife.com, 20 September 2023.

12 'Greenhouse Gas Emissions – Thursday, 26 Apr 2018 – Parliamentary Questions (32nd Dáil)', Houses of the Oireachtas, 26 April 2018.

13 John Gibbons, 'Ireland's Government "Using Fake Data to Pretend Dairy Emissions Aren't Rising Fast"', DeSmog.com, 25 June 2018.

14 Josh Gabbatiss, 'Irish Government Using Wrong Data to Downplay Greenhouse Gas Emissions from Cows', *Independent*, 26 June 2018.

15 'Opening Statement by Bill Callanan, Chief Inspector, DAFM, Prepared for Joint Committee on Environment and Climate Action, Regarding Recommendations Contained in Ireland's

Citizens' Assembly on Biodiversity Loss Relating to Land Use and Water Quality', Department of Agriculture, Food and the Marine, 7 November 2023.

16 'Joint Committee on Environment and Climate Action Debate – Micheál Ó Cinnéide', Houses of the Oireachtas, 19 September 2023.

17 Gary Murphy, 'The Irish Government, the National Farmers Association, and the European Economic Community, 1955–1964', *New Hibernia Review* 6(4), 2002: 68–84.

18 'Taoiseach's Stern Warning to N.F.A.: Association Threatened with Proscription', *Irish Times*, 25 April 1967.

19 Jack Fagan, 'Diary of a Nine-Month Struggle: N.F.A. and the Highlights', *Irish Times*, 31 May 1967.

20 Declan O'Brien, '50 Years in the EU: Farmers Lead on Road to EEC Membership', *Irish Farmers Journal*, 4 January 2023.

21 'IFA Launches "The Path to Power" – a 60-Year History of the Organisation', *Irish Farmers Journal*, 21 April 2015.

22 Eoin Burke-Kennedy, 'Smith Settlement Is a Difficult Pill to Swallow for IFA', *Irish Times*, 23 February 2018.

23 Horace Curzon Plunkett, *Ireland in the New Century* (New York, Dutton, 1908).

24 Patrick Doyle, *Civilising Rural Ireland: The Co-Operative Movement, Development and the Nation-State, 1889–1939* (Manchester University Press, 2019).

25 Proinnsias Breathnach, 'The Evolution of the Spatial Structure of the Irish Dairy Processing Industry', *Irish Geography* 33(1), 2000: 166–84.

26 'InFocus with Tara McCarthy', GrantThornton.ie, 19 December 2017.

27 'Food Vision 2030: Another Detrimental Blueprint for Agriculture', Environmental Pillar, 3 August 2021.

28 Environmental Protection Agency and Laura Burke, 'Re: Strategic Priorities to 2030 for "A Climate Smart, Environmentally Sustainable AgriFood Sector"', Environmental Protection Agency, 20 July 2020.

29 'Irish Examiner View: Revise Shameful Food Plan and Face Reality', Editorial, Irish Examiner, 7 August 2021.

30 'The Sunday Times View on Sustainable Farming: Ministers Have Shied Away from Unpalatable Truths about Food', Editorial, Sunday Times, 7 August 2021.

31 'Methane – What Is Methane (CH_4)?', Teagasc, Agriculture and Food Development Authority, 2024.

32 Joey Grostern, 'EU Nitrates Directive', DeSmog.com, November 2024.

33 Wim De Vries, 'Impacts of Nitrogen Emissions on Ecosystems and Human Health: A Mini Review', Current Opinion in Environmental Science & Health 21, June 2021: 100249; 'Dairy – Greenhouse Gas Mitigation Potential of Chemical N Fertiliser', Teagasc, December 2022.

34 'Commission Rules Out Re-Visiting Ireland's Nitrates Derogation Decision', Teagasc, 6 September 2023.

35 Amy Forde, 'Why Denmark Won't Be Renewing Its Nitrates Derogation', Irish Farmers Journal, 18 April 2024.

36 Ryan Harmon, 'Slurry: The Risks and Necessary Precautions', Farm Safely, 7 July 2022.

37 Andrew Hamilton, 'Illegal Slurry Spreading Must Be Stopped with Zero Tolerance Policy, Says Charlie McConalogue', Irish Independent, 31 October 2023.

38 Niall Hurson, '"Fraudulent Movements" Widespread in Slurry Export System', Irish Independent, 30 January 2024.

39 'IFA Will Reject Any Proposal to Impose EPA Licensing on Dairy Farmers', Irish Farmers' Association, 11 November 2020.

40 Eoghan Dalton, 'Dozens More Inspectors Needed to Tackle the "Wicked Problem" of Farm Water Pollution', TheJournal.ie, 18 June 2023.

41 Rubina Freiberg, 'Farm Inspection Rates "Far Below" Level Required to Improve Water Quality – EPA', Agriland.ie, 3 December 2024.

42 Brian H. Jacobsen et al., 'Health-Economic Valuation of Lowering Nitrate Standards in Drinking Water Related to Colorectal Cancer in Denmark', *Science of the Total Environment* 906, 2024: 167368.

43 Amy Forde, 'Why Denmark Won't Be Renewing Its Nitrates Derogation', *Irish Farmers Journal*, 18 April 2024.

44 'Summary Report: Water Quality in Ireland, 2016–2021', Environmental Protection Agency, October 2022.

45 Francess McDonnell, 'Agriculture Run-Offs to Blame for Decline in Water Quality: EPA', Agriland.ie, 14 October 2022.

46 Ciaran Moran, '"Job of Improving Water Quality Much Harder Than It Was 30 Years Ago" – Teagasc', *Irish Independent*, 14 May 2024.

47 Brendan O'Connor et al., 'Coastal Lagoons: Ecology and Restoration (CLEAR)', EPA Research Report no. 473, Environmental Protection Agency, 2025.

48 Lorna Siggins, 'Polluted Irish Lagoon That Can Be Seen Glowing from Space Could Cost Millions to Clean Up', *Irish Independent*, 19 January 2025.

49 'Ireland's Provisional Greenhouse Gas Emissions 1990–2023', Environmental Protection Agency, July 2024.

50 Emma Gilsenan, '"Believe It or Not We Never Considered the Outcome of the Dairy Calves" – Teagasc', Agriland.ie, 22 October 2019.

51 Ciaran Moran and Niall Hurson, 'Dairy Sector Braces for Fallout from RTÉ Investigates "Milking It: Dairy's Dirty Secret" Exposé', *Irish Independent*, 11 July 2023.

52 'The Teagasc Authority', February 2024, Teagasc website.

53 Adam Woods, 'Row Erupts Over Teagasc Advice to Switch to Dairy Beef', *Irish Farmers Journal*, 22 September 2021.

54 FAB Team, 'Opposition from Irish Dairy Sector Against Proposed Dairy Exit Program and Voluntary Cow Culling', Food and Beverage Business (blog), 15 August 2023; Hannah Quinn Mulligan, 'Government Considers €2bn Farm Retirement Scheme to Meet Climate Targets', *Irish Times*, 22 September 2022.

55 'Executive Summary: Marginal Abatement Cost Curve 2023', Teagasc, 12 July 2023.

56 Eoin Burke-Kennedy, '"Factory Levy" Generates up to Half of IFA Income', *Irish Times*, 24 November 2015.

57 'Farm Business – Farm Incomes Move up in 2024', Teagasc, 3 December 2024.

58 Oonagh Smyth, 'Priced Out: Concern Rising Land Costs Could Alter Face of Farming', RTÉ News, 22 August 2024.

59 Brendan Kearney, 'Small Farmers Have No Business Being in Beef', *Irish Times*, 28 November 2019.

60 Anniek J. Kortleve et al., 'Over 80 per cent of the European Union's Common Agricultural Policy Supports Emissions-Intensive Animal Products', *Nature Food* 5(4), April 2024: 288–92.

61 'EU Agricultural Spending Has Not Made Farming More Climate Friendly' (press release), European Court of Auditors, 21 June 2021.

62 'Towards EU Climate Neutrality Progress, Policy Gaps and Opportunities: Assessment Report 2024', European Scientific Advisory Board on Climate Change, 17 January 2024.

63 Amy Forde, 'Scrap VAT Exemptions on Fertiliser and Feed – OECD Report', *Irish Farmers Journal*, 12 May 2021.

Chapter 3: Powering Our Future

1 'ESB History Timeline', ESB Corporate, 2025.

2 Michael Shiel, *The Quiet Revolution: The Electrification of Rural Ireland, 1946–1976* (O'Brien Press, 2003).

3 John Gibbons, 'Answer to Our Energy Needs Is Blowing in the Wind', *Irish Times*, 5 March 2009.

4 Anthea Lacchia, '"Death by a Thousand Cuts": Hydropower Killing, Injuring and Trapping Fish by the Tonne', TheJournal.ie, 7 March 2023.

5 'Share of Energy from Renewable Sources', Eurostat, 2024.

6 Kevin O'Sullivan, 'State Needs to Stop Spending €1m an Hour Importing Fossil Fuels, Conference Told', *Irish Times*, 22 September 2022.

7 'How Will Wind Energy Help Ireland Reach Its Climate Targets and Reduce Its Emissions?', Wind Energy Ireland, 2022.

8 Paul O'Donoghue, 'Why Ireland's Vision of Being the "Saudi Arabia of Offshore Wind" Is in Trouble', TheJournal.ie, 10 August 2024.

9 'Growth of Onshore to Offshore Wind – Atlantic Region Wind Energy & Supply-Chain Feasibility Study', Dublin Offshore, July 2022.

10 'New Report Identifies Ireland's €38 Billion Offshore Wind Opportunity', Wind Energy Ireland, 29 January 2024.

11 'Future Framework for Offshore Renewable Energy: Policy Statement 2024', Government of Ireland, 2024.

12 'Minister Ryan Welcomes Hugely Positive Provisional Results of First Offshore Wind Auction', Department of the Environment, Climate and Communications, 11 May 2023.

13 'Codling Wind Park', Codling Wind Park, 2024.

14 'Wind Energy Saved Ireland over €1.2 Billion on Gas in 2024', Wind Energy Ireland, 16 February 2025.

15 Barry O'Halloran, 'Equinor Withdrawal Puts Focus on Off-shore Planning Delays', *Irish Times*, 4 November 2021.

16 Paul Deane, interviewed by the author, July 2024.

17 J. Humphreys, '*Coolglass Wind Farm Limited and An Bord Pleanála (Applicants)* vs. *Ireland and the Attorney General (Respondent)* – High Court Judgment', 10 January 2025.

18 Eoin Burke-Kennedy, '€2bn Dublin Bay Wind Farm to Submit Planning Application', *Irish Times*, 26 February 2025.

19 James MacDonald, 'How Offshore Oil Exploration Affects Marine Life', JSTOR Daily, 1 February 2019.

20 'Denmark – a Frontrunner in Wind Energy', State of Green (blog), 17 November 2021.

21 'Vestas Wins 56 MW Order in Ireland', Vestas, 20 December 2022.

22 'Wind Energy in Denmark', International Energy Agency Wind Technology Collaboration Programme, 2023.

23 'Danish Energy Transition "Stalled" After 3GW Auction Flop', *reNEWS – Renewable Energy News*, 6 December 2024.

24 'Offshore Wind Turbines Part of Danish Touristic Offer', Offshore Wind (blog), 17 November 2016.

25 Bob Berwyn, 'How Do Wind Farms Affect Ocean Ecosystems?', dw.com, 22 November 2017.

26 Andreas Tang, 'Energy Islands Coming to Europe's Seas', windeurope.org, 14 November 2022.

27 Naomi O'Leary, 'Ireland Joins European Project to Scale Up Offshore Wind Energy', *Irish Times*, 24 April 2023.

28 'Powering Prosperity – Ireland's Offshore Wind Industrial Strategy', Department of Enterprise, Trade and Employment, 8 March 2024.

29 'Ports Policy – Thursday, 4 Jul 2024 – Parliamentary Questions (33rd Dáil)', Houses of the Oireachtas, 4 July 2024.

30 'More about COER', Centre for Ocean Energy Research (blog), 26 November 2020.

31 Zerina Maksumic, 'Wave Energy Can Meet Double Ireland's Power Needs – If Supported, COER Says', Offshore Energy (blog), 14 November 2024.

32 'Key Findings: Data Centres Metered Electricity Consumption 2023', Central Statistics Office, 23 July 2024.

33 Ian Curran, 'Data Centres Consume as Much Electricity as Urban Houses, CSO Figures Show', *Irish Times*, 12 June 2023.

34 'Government Statement on the Role of Data Centres in Ireland's Enterprise Strategy', Department of Business, Enterprise and Innovation, June 2018.

35 Heather Rogers, 'Current Thinking', *New York Times* magazine, 3 June 2007.

36 'In 2004, It Took the World a Year to Add a Gigawatt of Solar Power – Now It Takes a Day', Our World in Data, 7 February 2025.

37 Will Goodbody, '1GW of Solar Capacity Now Connected to Electricity Grid', RTÉ News, 26 February 2024.

38 'ESB Networks Announces 100,000 Microgenerators Are Now Connected to Ireland's Electricity Network', ESB Corporate, 12 July 2024.

39 'Energy in Ireland: 2024 Report', Sustainable Energy Authority of Ireland, December 2024.

40 Eithne Dodd, 'Growing Interest Among Farmers Obtaining Solar Panels', RTÉ News, 6 March 2024.

41 Noel Bardon, 'Up to €1,200/Ac to Lease Farmland to Solar Developers', *Irish Farmers Journal*, 2 August 2023.

42 '"UK Solar Farms Can Be Wildlife Havens"', *reNEWS – Renewable Energy News*, 23 May 2023.

43 Matthew Eisenson, 'Solar Panels Reduce CO_2 Emissions More Per Acre Than Trees – and Much More Than Corn

Ethanol – State of the Planet', Climate Law Blog, Columbia University's Sabin Center for Climate Change Law, 26 October 2022.

44 Ciaran Moran, '"It Could Lead to Devastating Consequences for the Dairy Industry" – Massive Cork Solar Farm on Top Dairy Site Criticised by Local TD', *Irish Independent*, 9 April 2025.

45 'Solar Panels Ireland Cost: 2025 Guide', Energy Efficiency Ireland (blog), 31 May 2023.

46 'Microgeneration for Homes', Electric Ireland, 2025.

47 Gráinne Ní Aodha, 'Michael O'Leary "Astonished" by Green Power of His Farm's Solar Panels', BreakingNews.ie, 7 October 2024.

48 Alex Lawson, 'France to Require All Large Car Parks to Be Covered by Solar Panels', *Guardian*, 9 November 2022.

49 Myles Buchanan, 'Wicklow County Council's Solar Car Port Project Showcased at National Seminar', *Wicklow People*, 31 October 2023.

50 '40 Years: Turlough Hill', ESB, 2014.

51 'Minister Ryan Outlines Ambitious Electricity Interconnection Plans as Ireland Aims to Become Net Energy Exporter', Department of the Environment, Climate and Communications, 26 July 2023.

52 'Interconnection', System Operator for Northern Ireland soni.ltd.uk/industry/interconnection, 2025.

53 Andy Colthorpe, 'Electricity Supply Board Opens Ireland's Largest Battery Storage Facility at Dublin Energy Hub', *Energy-Storage News*, 13 February 2024.

54 Marija Maisch, 'Ireland in Line for 1 GWh Iron-Air Battery Storage Project', *PV Magazine International*, 27 September 2024.

55 'Kestrel Project', h2kestrel.ie.

56 'The Problem with Hydrogen', Global Witness, 1 September 2022.

57 Linda Daly, 'Bord Gais Boss Engineers a Transition to Green Energy', *Sunday Times*, 29 June 2024.

58 Peter Harte, 'The Role of Long Duration Energy Storage (LDES) in Delivering a 100 per cent Decarbonised Power System', Engineers Ireland webinar video, 2024.

59 'Agriculture's Role in Biomethane Production', Teagasc, 6 May 2023.

60 Cathal G. Nolan et al., 'Risk of Drought-Related "Fodder Crises" in Irish Agriculture Under Mid-21st Century Climatic Conditions', National Hydrology Conference, November 2021.

61 Niall Hurson, 'Almost Two-Thirds of Dairy Farmers Considering Cull Due to Fodder Shortage – ICMSA Survey', *Irish Independent*, 16 July 2024.

62 John Carey, 'While Some Tout "Renewable Natural Gas" as a Way to Mitigate Climate Change, Others See a False Solution', *Proceedings of the National Academy of Sciences* 120(28), 2023: e2309976120.

63 Tim Searchinger and Ralph Heimlich, 'Avoiding Bioenergy Competition for Food Crops and Land', World Resources Institute, 28 January 2015.

64 Eamon Ryan, 'Storm Éowyn Is a Reminder Why We Need More EVs and Fewer Sitka Spruce Trees', *Irish Times*, 11 February 2025.

65 Kevin O'Sullivan, 'After Storm Éowyn: Can Ireland's Electricity, Water and Phone Networks Cope with Extreme Weather?', *Irish Times*, 31 January 2025.

66 Steve Sorrell, 'Jevons Paradox Revisited: The Evidence for Backfire from Improved Energy Efficiency', *Energy Policy* 37(4), 2009: 1456–69.

67 William Walsh, 'Tough Choices Must Be Made on the Climate Crisis', *Irish Times*, 28 January 2025.

68 Daniel Murray, 'Government Approves Ireland's First LNG Facility in €300m "State-Led" Project', *Business Post*, 4 March 2025.

Chapter 4: Ireland on the Move

1 'Transport Emissions, 1990–2023', Environmental Protection Agency, 1 May 2024.

2 Seán McCárthaigh, 'Irish People Have the Second Highest Level of Car Dependency Among EU Citizens', *Irish Independent*, 16 October 2022.

3 Andrew Thacker, *Moving Through Modernity: Space and Geography in Modernism* (Manchester University Press, 2009), p. 127.

4 'Transport Trends 2021 – An Overview of Ireland's Transport Sector', Department of Transport, 5 July 2022.

5 Eddie Cunningham, 'Revealed: How Our Cars Are Getting Heavier by the Year', *Irish Independent*, 3 July 2024.

6 'Chapter 11: Environment and Transport', State of the Environment Report, Environmental Protection Agency, July 2024.

7 'National Household Travel Survey 2023 Research Report', National Transport Authority, December 2023, p. 19.

8 Seán O'Riordan, 'No "Silver Bullet" to Solve Congestion at Jack Lynch Tunnel, According to TII', *Irish Examiner*, 14 July 2024.

9 Barry Roche, 'Cork–Limerick Road to Be Upgraded to Full Tolled Motorway at Estimated Cost of €2bn', *Irish Times*, 24 January 2024.

10 'All-Island Strategic Rail Review – Final Report', Department of Transport and Department for Infrastructure, 31 July 2024.

11 Christina Finn, '"The People Want Tar": Ministers Herald "New Dawn" of Increased Roads Spending Worth €713m', TheJournal.ie, 14 February 2025.

12 'Draft Dublin City Centre Transport Plan', Dublin City Council, 7 February 2024.

13 'Ibec Calls for Immediate Pause on Dublin City Traffic Plans', Ibec, 20 May 2024.

14 Darragh Moriarty, 'Ibec's Disingenuous Last Ditch Attempt to Derail Dublin Transport Plan Must Be Rejected', Irish Labour Party news release, 22 May 2024.

15 Olivia Kelly, 'Traffic Plan Will Cost Dublin €400m a Year, Says Report for Traders Opposed to Car Curbs', *Irish Times*, 11 July 2024.

16 Daniel McConnell, 'Brown Thomas Arnotts Boss Voices "Deep Concerns" Over Dublin City Transport Plan', *Business Post*, 11 July 2024.

17 Donal Macnamee, '"Electric": Brown Thomas Arnotts' Boss Hails Strong Christmas', *Business Post*, 29 December 2024.

18 'Cost of Motoring 2019', AA Ireland annual survey.

19 'Census 2022 Profile 7 – Employment, Occupations and Commuting', Central Statistics Office, 5 December 2023.

20 James Paskins, 'Understanding the Car Dependency Impacts of Children's Car Use', paper written for the workshop 'Children and Traffic', Copenhagen, 2 May 2002.

21 Mairead Cleary, 'One in Five Primary School Children Are Obese or Overweight, According to HSE Survey', *Irish Examiner*, 14 October 2020.

22 Lawrence D. Frank, 'Obesity Relationships with Community Design, Physical Activity, and Time Spent in Cars', *American Journal of Preventive Medicine* 27(2), 2004: 87–96.

23 'Redesigning Ireland's Transport for Net Zero – Towards Systems that Work for People and the Planet', OECD, 5 October 2022.

24 'Our Network: About ESB ecars', ESB, 2023.

25 Natalie Middleton, 'Myths and Misinformation About EVs Now 'Deeply Embedded', *Fleetworld*, March 2024.

26 David Reichmuth, 'Driving Cleaner – Electric Cars and Pickups Beat Gasoline on Lifetime Global Warming Emissions', Union of Concerned Scientists, July 2022.

27 Jasper Jolly, 'Do Electric Cars Pose a Greater Fire Risk Than Petrol or Diesel Vehicles?', *Guardian*, 20 November 2023.

28 Conor Molloy, 'Understanding EV Battery Life', Sustainable Energy Authority of Ireland, 24 February 2023.

29 'Range Therapy – Generate Power, Not Pollution', Range Therapy corporate website.

30 'EVs to Surpass Two-Thirds of Global Car Sales by 2030, Putting at Risk Nearly Half of Oil Demand, New Research Finds', Rocky Mountain Institute, 14 September 2023.

31 Online EV journey cost calculator, Sustainable Energy Authority of Ireland.

32 'Annual Review 2024 – Transport', Climate Change Advisory Council, 27 June 2024, p. 12.

33 Max Baumhefner, 'How Electric Cars and Trucks Improve Grid Reliability', Natural Resources Defense Council, 8 September 2022.

34 'Ever-Wider: Why Large SUVs Don't Fit, and What to Do About It', Transport & Environment, 22 January 2024.

35 Ian Walker, Alan Tapp and Adrian Davis, 'Motonormativity: How Social Norms Hide a Major Public Health Hazard', *International Journal of Environment and Health* 11(1), 2023: 21–33.

36 Courtney Coughenour et al., 'Estimated Car Cost as a Predictor of Driver Yielding Behaviours for Pedestrians', *Journal of Transport & Health* 16, 2020: 100831.

37 B. Claus and L. Warlop, 'Car Cushion Hypothesis: Bigger Cars Lead to More Risk Taking – Evidence from Behavioural Data', *Journal of Consumer Policy* 45, 2022: 331–42.

38 'Flashback: The Pedestrianisation of Grafton Street Began 45 Years Ago Today', *Irish Independent*, 5 September 2016.

39 Ger Siggins, 'Flashback 1982: Closing of Grafton Street to Traffic', *Irish Independent*, 29 November 2015.

40 Ellen O'Regan, 'Cork's Outdoor Dining and Pedestrianisation Experiment to Continue', *Irish Examiner*, 23 May 2023.

41 Angelique Chrisafis, 'Parisians Vote in Favour of Tripling Parking Costs for SUVs', *Guardian*, 4 February 2024.

42 Lauren Boland, 'Bonjour to Bikes: Cycling More Popular Than Driving in Paris City Centre', TheJournal.ie, 11 April 2024.

43 Aitor Hernández-Morales, 'The City that Pioneered Europe's Car-Free Future', *Politico*, 27 July 2022.

44 'Investing in Pedestrian Areas Multiplies Local Income', BBVA Private Banking, 2022.

45 Ibid.

46 Muiris O'Cearbhaill, 'Regina Doherty: "I worded it clumsily the first time. But authoritarian cycle lanes have divided towns"', TheJournal.ie, 17 July 2024.

47 'Redesigning Ireland's Transport for Net Zero – Towards Systems that Work for People and the Planet', OECD, 5 October 2022, p. 87.

48 Ian Curran, 'Permanent TSB Replaces Renault as Late Late Show Sponsor', *Irish Times*, 15 September 2023.

49 Ellen Coyne, 'Deals on Wheels: The RTÉ Stars Driving Cars Worth Up to €70,000 Thanks to Lucrative Brand Tie-Ups', *Irish Independent*, 1 July 2023.

50 Lottie Limb, 'French Car Adverts Must Encourage People to Bike and Walk', EuroNews, 7 January 2022.

51 David Vetter, 'This Scottish City Just Banned SUV and Airline Ads. Here's Why', *Forbes*, 30 May 2024.

52 Shane Timmons et al., 'Active Travel Infrastructure Design and Implementation: Insights from Behavioural Science', *WIREs Climate Change* 15(3), 2024: e878.

53 'What Stops Women Cycling in London?', London Cycling Campaign Women's Network Report, 2024.

54 Michael Sheridan, 'Toyota Aygo X: Composed and Impressive for Its Class', *Irish Times*, 8 February 2022.

55 Seán Owens, 'Why Are We in Such a Hurry to Kill Ourselves . . . and the Planet?', Dundalk Cycling Alliance, October 2024.

56 'Why Your Journey Counts', Department of Transport, 17 May 2024.

57 Matt Walsh, 'Phase 6a of BusConnects Will Launch on January 26th 2025', Dublin Public Transport, 21 January 2025.

58 'All Ireland Rail Review Proposes €37 billion Transformation', Rhomberg Sersa Rail Group Ireland, 1 August 2023.

59 'What is DART+ Programme?', Dartplus.ie.

60 'Ireland's Road Haulage Strategy 2022–2031', Department of Transport, December 2022.

61 Tarun Joshi, 'Are Railroads the Most Environmentally Friendly Solution in Freight Transportation?', Responsible Investment Association, 17 November 2022.

62 'Modal Split of Inland Freight Transport – Railways', Eurostat, 15 April 2024.

63 Hannah Daly, 'Carbon Emissions from Aviation Can't Be Swept Under the Carpet', *Irish Times*, 6 June 2024.

64 'Excise Duties on Energy – EU Legislation on Excise Duties for Energy Products', European Commission.

65 'Budget 2023 Continues to Keep Costs Down for Commuters and Ensures that Transport Developments Can Continue at Pace', Department of Transport, 27 September 2022.

66 'VAT Treatment of Services Relating to Vessels and Aircraft', Revenue.ie, October 2024.

67 Sourasis Bose, 'US Sustainable Aviation Fuel Production Target Faces Cost, Margin Challenges', Reuters, 1 November 2023.

68 '1 per cent 'Super Emitters' Responsible for Over 50 per cent of Aviation Emissions', Transport & Environment, 3 December 2020.

69 'European Private Jet Pollution Doubled in One Year', Greenpeace European Unit, 20 March 2023.

70 Daniel Murray, 'Dublin Airport Emissions Could Rise by 22 Per Cent if Passenger Cap Lifted, DAA Document Says', *Business Post*, 8 February 2024.

71 Órla Ryan, 'I'm not rejecting science, I'm supporting the economy': DAA Boss Defends Plan to Increase Flights', TheJournal.ie, 7 February 2024.

72 'Ireland's Provisional Greenhouse Gas Emissions 1990–2023', Environmental Protection Agency, July 2024, p. 31.

Chapter 5: The New Land War

1 'Water Quality Monitoring Report: Discussion', Joint Oireachtas Committee on Agriculture, Food and the Marine, 19 July 2023.

2 Mairead Maguire, 'Government Using Farmers as "Scapegoats" for Increase in Emissions – McNamara', Newstalk, 12 February 2023.

3 Brian Mahon, '"Worse than Cromwell" – Ireland Must Cut Livestock Numbers by a Third', Extra.ie, 11 February 2023.

4   ''I will never apologise to your like." Rural TD Defends His "Inflammatory" Comments on Herd Culling', Virgin Media Television, June 2023.

5   J. J. Byrne, 'Agriculture: General Survey', in James Meenan and David A. Webb (eds.), *A View of Ireland: Twelve Essays on Different Aspects of Irish Life and the Irish Countryside* (British Association for the Advancement of Science, 1957); 'Farm Structure Survey, 2023', Central Statistics Office, 17 December 2024.

6   'State of the World's Birds', Birdlife International, 31 January 2024.

7   'The World is Hungry for Food Sustainability', promotional video for Origin Green, Bord Bia, 2012.

8   'Bord Bia to Invest €3.5 Million Promoting Ireland's Sustainable Food Industry', Bord Bia, 11 December 2013.

9   'Bord Bia Board Members', Bord Bia, 2024.

10 Catherine Shanahan, 'Eight Firms Placed on EPA Watchlist', *Irish Examiner*, 31 January 2018.

11 'EPA Prosecutes Dawn Meats Ireland Unlimited Company', Environmental Protection Agency, 15 May 2023.

12 'About Bord Bia – Our Story', Bord Bia.

13 Sarah Harford, 'The Irish Food Board Wants to Know How to "Win Back" Vegetarians and Vegans', TheJournal.ie, 2 June 2018.

14 'Bord Bia Launches New EU Beef, Lamb & Dairy Promotions Worth €8m', Bord Bia, 1 September 2022.

15 Aisling O'Brien, 'Premium Price for Irish Beef Flown Weekly to Singapore', Agriland.ie, 4 September 2022.

16 David Raleigh, 'Large Shipment of Ukrainian Grain to Land in Ireland – First Since War Began', *Irish Examiner*, 20 August 2022.

17 'Marginal Abatement Cost Curve 2023', Teagasc, 12 July 2023.

18 'Calculation of Methane Emissions: Discussion', Joint Oireachtas Committee on Agriculture, Food and the Marine, 20 July 2022.

19 'Setting The Record Straight – Prof. Frank Mitloehner', *Countrywide*, RTÉ Radio 1, 21 January 2023.

20 Zach Boren, 'Revealed: How the Livestock Industry Funds the "Greenhouse Gas Guru"', Unearthed, Greenpeace, 31 October 2022.

21 Anne O'Donoghue, 'In Full: Farming Commitments in the Programme for Government', *Irish Farmers Journal*, 15 January 2025.

22 'The Dublin Declaration of Scientists on the Societal Role of Livestock', Dublin-declaration.org, 2022.

23 'Leading Scientists Advance Science in Global Meat Debate', Teagasc News, 19 October 2022.

24 David Murphy, 'Teagasc Criticised for Advocating "Societal Role" of Meat', RTÉ News, 31 December 2024.

25 Zach Boren, 'Revealed: The Livestock Consultants Behind the Dublin Declaration of Scientists', Unearthed, Greenpeace, 27 October 2023.

26 'The Dublin Declaration – Signatories', Dublin-declaration. org.

27 Zach Boren, 'Revealed: The Livestock Consultants Behind the Dublin Declaration of Scientists', Unearthed, Greenpeace, 27 October 2023.

28 'The Dublin Declaration – Authorship', Dublin-declaration. org.

29 Zach Boren, 'Revealed: The Livestock Consultants Behind the Dublin Declaration of Scientists', Unearthed, Greenpeace, 27 October 2023.

30 Ibid.

31 'The Role of Meat in Society: Presenting the Dublin Declaration of Scientists', FEFAC, 12 April 2023; 'Animal Task Force, Belgium Association of Meat Science & Technology Symposium', YouTube, April 2023.

32 Jochen Krattenmacher et al., 'The Dublin Declaration: Gain for the Meat Industry, Loss for Science', *Environmental Science & Policy* 162, December 2024: 103922.

33 '2023 Annual Report & Financial Statements', Teagasc, 27 September 2024.

34 Geraldine Tallon et al., 'Independent Evaluation of the Climate Change Advisory Council', CCAC, December 2020.

35 Killian Flood, 'Supreme Court: Environmental Challenge to Glanbia Cheese Factory Dismissed', *Irish Legal News*, 17 February 2022.

36 David Condon, 'Jackie Cahill Slams An Taisce's Kilkenny Cheese Plant Appeal as "a revolting act of treason"', Tipp Mid West Radio, 10 May 2021.

37 Philip Ryan, 'Two TDs Who Criticised An Taisce Objections Have Shares in Glanbia', *Irish Independent*, 14 May 2021.

38 Claire McCormack and Declan O'Brien, 'Pressure Mounts in Dairy Sector as TDs Weigh in on Glanbia Row', *Irish Independent*, 6 April 2021.

39 Sarah Stack, 'Glanbia Top Executives Served Up Multi-Million Euro Package', *Irish Independent*, 13 July 2014.

40 Eoin Burke-Kennedy, 'Glanbia Boss Siobhán Talbot Sees Salary Jump 73 Per Cent to Nearly €6m', *Irish Times*, 10 March 2023.

41 Eoghan Dalton, 'Tirlán Boss Wants Govt to Give €40m to Plan Aimed at Improving "Worst Rivers" in Ireland', The-Journal.ie, 20 April 2024.

42 Aidan Brennan, 'Bring Forward Climate and Nature Fund, Says Tirlán CEO', *Irish Farmers Journal*, 8 May 2024.

43 'ICSA Condemns the Trolling of Livestock Farmers', *Wexford People*, 22 February 2023.

44 Páraic McMahon, 'Clare Farmers Annoyed by "Hippy Dippies & Tree Huggers" and Concerned for Future of Sector', *Clare Echo*, 9 October 2021.

45 Colin Gleeson, 'Dairy Farmers Criticise "Attack" by "Arrogant" Environmentalists', *Irish Times*, 22 January 2020.

46 Barry Murphy, 'Environmental Groups "Holding the National Interest Hostage" – ICMSA', *Irish Farmers Journal*, 3 March 2022.

47 'Reaction to EU's Nature Restoration Law', *Drivetime*, RTÉ Radio 1, 15 June 2023.

48 Kevin O'Sullivan, 'EU Official Castigates Government Over Environmental Court Costs', *Irish Times*, 21 January 2022.

49 Zoe Kavanagh, 'Trust Issues', *Dairy Ireland*, 2022, pp. 8–9.

50 Kathleen O'Sullivan, 'Complaint Against NDC Claim of Irish Dairy's "Most Emissions-Efficient Production System" Upheld', *Irish Examiner*, 21 December 2023.

51 Niall Hurson, 'Farm Groups and Agri-Business Stakeholders' Plan Emerges to Team Up to Tackle Farming's Negative Image', *Irish Independent*, 7 May 2024.

52 Ibid.

53 'Climate Change in the Irish Mind – Wave 2', Environmental Protection Agency, 2024.

54 'Majority Underestimate Climate Impact of Food', Economic & Social Research Institute, 9 May 2024.

55 Caroline O'Doherty, 'Creator of Kim Cattrall "Meat Tweet" Quit Her Job After Climb-Down by EPA Over Furious Farmers', *Irish Independent*, 29 August 2024.

56 Marie O'Halloran, 'Rural TDs Criticise Varadkar for Comments on Eating Meat', *Irish Times*, 15 January 2019.

57 John Gibbons, 'How Big Ag Is Influencing What Irish Students Learn About Climate Change', DeSmog, 15 December 2020.

58 'Moo Crew – EU Milk Scheme', National Dairy Council.

59 'Research for AGRI Committee – Policy Support for Productivity vs Sustainability in EU Agriculture: Towards Viable Farming and Green Growth', European Parliament, January 2017.

60 Francess McDonnell, 'GHG Emissions Per Euro of Agri Output Have Fallen – Report', Agriland.ie, 5 December 2024.

61 Orla Dwyer, 'FactFind: Are Irish Farmers the "Most Carbon-Efficient Food Producers in the World"?', TheJournal.ie, 24 July 2022.

62 Thomas Pringle, 'School Curriculum', Dáil Éireann Debate, Houses of the Oireachtas 8 December 2020.

63 'Frequently Asked Questions – Animal Welfare', National Dairy Council website.

Chapter 6: Getting Our Houses in Order

1 'Residential Final Energy Demand', Sustainable Energy Authority of Ireland, 2022.

2 Pádraig Hoare, 'Vast Majority of New Homes Meet Highest Energy Standards', *Irish Examiner*, 17 January 2022.

3 'Statistics for National Home Retrofit Programmes', Sustainable Energy Authority of Ireland, 31 January 2025.

4 Eoin Burke-Kennedy, 'More Than Two-Thirds of People Are Living in Homes Too Big for Their Needs, ESRI Research Finds', *Irish Times*, 27 March 2024.

5 'Housing in Europe – 2023 Edition', Eurostat, November 2023.

6 'Domestic Building Energy Ratings, Quarter 4 2024', Central Statistics Office, 21 January 2025.

7 Olivia Kelly, 'Ireland Is Building Too Many Large Detached Houses, Says Construction Body', *Irish Times*, 28 November 2022.

8 'Retrofit – Costs and Fees', Electric Ireland Superhomes.

9 'New Low-Cost Home Energy Upgrade Loan Scheme Launched', Sustainable Energy Authority of Ireland, 24 April 2024.

10 Marie Hyland et al., 'The Value of Domestic Building Energy Efficiency – Evidence from Ireland', Economic & Social Research Institute research bulletin, 1 March 2014; Ciaran Byrne, 'The Value of a Home Energy Upgrade', Sustainable Energy Authority of Ireland (blog), 4 July 2024.

11 'Government Supported 47,900 Home Energy Upgrades Through SEAI in 2023', Sustainable Energy Authority of Ireland, 7 March 2024.

12 Dominic Ó Gallachóir, 'Heat Pump vs Oil Boiler Calculator', IrishHeatPumps.com.

13 Phoebe Cooke, 'Revealed: Media Blitz Against Heat Pumps Funded by Gas Lobby Group', DeSmog, 20 July 2023.

14 'Gas Networks Ireland Outlines a Pathway to a Net Zero Carbon Network by 2045', Gas Networks Ireland, 26 June 2024.

15 Jan Rosenow, 'Is Heating Homes With Hydrogen All But a Pipe Dream? An Evidence Review', *Joule* 6(10), 2022: 2225–8.

16 'Climate Action Plan 2024', Government of Ireland, 2 July 2024, p. 206.

17 'Accelerating the Energy Efficiency Renovation of Residential Buildings: A Behavioural Approach', European Environment Agency, 29 June 2023.

18 'How District Heating Is Paving the Way Towards Denmark's Climate Goals', State of Green (blog), 6 November 2023.

19 'Climate Action Plan 2024', Government of Ireland, 2 July 2024, p. 219.

20 'Minister Ryan Launches Report Paving the Way for the Expansion of District Heating to Irish Homes and Businesses by 2030', Department of the Environment, Climate and Communications, 31 August 2023.

21 'Report on the Dublin District Heating Project – Dublin City Council', Dublin City Council, 5 May 2021.

22 'DDHP Initiative to Deliver the Largest Sustainable District Heating System in Ireland', Innovation News Network, 16 October 2024.

23 'Annual Report 2023 on Public Sector Energy Performance', Sustainable Energy Authority of Ireland, 2023, p. 25.

24 Alan Barrett and John Curtis, 'The National Development Plan in 2023: Priorities and Capacity', Economic & Social Research Institute, 12 January 2024, p. 37.

25 John FitzGerald, 'We Won't Fix the Housing Shortage Without Solving the Skills Shortage', *Irish Times*, 19 January 2024.

26 'Census of Population 2022 – Profile 2 Housing in Ireland, Central Statistics Office', 27 July 2023.

27 'Public Consultation on Removing Barriers to Energy Efficiency in the Rental Sector by Addressing the "Split Incentive"', Department of Communications, Climate Action and Environment, 16 January 2020.

28 Rory Hearne, *Gaffs: Why No One Can Get a House, and What We Can Do About It* (HarperCollins, 2022), p. 289.

29 'Micro-Generation – Clean Export Guarantee', Department of the Environment, Climate and Communications, 13 July 2021; Henry Fox, 'What Feed-in Tariffs Mean for Solar in Ireland', *Irish Tech News*, 18 July 2022.

30 'Sustainable Rural Housing – Guidelines for Planning Authorities', Department of Environment, Heritage and Local Government, 2005.

31 'Census of Population 2022 – Profile 2 Housing in Ireland', Central Statistics Office, 27 July 2023.

32 Ankita Gaur et al., 'Dispersed Settlement Patterns Can Hinder the Net-Zero Transition: Evidence from Ireland', *Energy Strategy Reviews* 51, January 2024: 101296.

33 'Oireachtas Joint Committee on Transport and Communications Debates', Houses of the Oireachtas, 21 April 2015, p. 31.

34 'Draft First Revision to the National Planning Framework', Government of Ireland, July 2024.

35 'Programme for Government 2025 – Securing Ireland's Future', Government of Ireland, 23 January 2025, p. 47.

36 Muhammad Ali et al., 'This Is Not Climate Resilience, It Is Suffering: How Storm Éowyn Tore a Hole in Our Preparations for Extreme Weather', *Irish Times*, 1 February 2025.

37 'Census 2022 and Vacant Dwellings FAQ', Central Statistics Office, 27 July 2023; Christina Finn, 'Census Shows 166,000 Vacant Properties in Ireland, With Over 48,000 Vacant for Six Years', TheJournal.ie, 23 June 2022.

38 'Annual Review 2024 – Built Environment', Climate Change Advisory Council, 10 July 2024, p. 8.

39 'Vacant Property Refurbishment Grant', Department of Housing, Local Government and Heritage, 19 September 2022.

40 Valerie Mulvin, *Approximate Formality: Morphology of Irish Towns* (Anne Street Press, 2021).

41 Michael Moynihan, 'Waterford Greenway Gives County "a destination status we've never had"', *Irish Examiner*, 20 November 2022.

42 Gemma Tipton, 'Reviving Small Irish Towns: "They are exactly the same as some of the finest European towns. They can become vibrant communities again"', *Irish Times*, 11 August 2024.

Chapter 7: A Climate of Doubt and Disbelief

1 David Robbins et al., *Ireland and the Climate Crisis* (Palgrave Macmillan, 2020).

2 Emmet Fox and Henrike Rau, 'Climate Change Communication in Ireland', *Oxford Research Encyclopaedia of Climate Science*, October 2016.

3 John Gibbons, 'As Climate Issues Intensify the Media, Incredibly, Throws in the Towel', *Irish Times*, 19 January 2012.

4 Olivia Ovenden, 'Adam McKay Is Still Trying to See the Funny Side in All This', *Esquire Magazine*, 13 November 2021.

5 Michael O'Leary, 'Climate Change and Cheap Flying', Letters, *Irish Times*, 12 July 2008.

6 John Gibbons, 'Bellamy "Late Late" Let-Off a Disservice to Climate Issue', *Irish Times*, 5 February 2009.

7 Ibid.

8 *The Pat Kenny Show*, RTÉ Radio 1, 16 November 2009.

9 John Gibbons, 'Kenny Stirs Up Bogus Climate Change Debate', *Irish Times*, 19 November 2009.

10 Bob Burton, 'Ian Plimer's Mining Connections', PR Watch, 12 November 2009.

11 John Gibbons, 'Climate Scientist Fillets Pat Kenny', Thinkorswim.ie (blog), 27 September 2011.

12 *The Pat Kenny Show*, Newstalk radio, 21 March 2023.

13 Mark Cullinane and Clare Watson, 'Irish Public Service Broadcasting and the Climate Change Challenge – Research Report and Findings', RTÉ Audience Council, February 2014.

14 'RTÉ Announces a Week of Programmes on Climate Change', RTÉ News, 24 October 2019.

15 Irish Doctors for the Environment, Open Letter to RTÉ, posted on Twitter, 22 July 2021.

16 Peter Thorne – Twitter posting, 12 July 2021.

17 Ciara O'Loughlin, 'RTÉ News Boss Apologises for Broadcaster Not Linking Recent Extreme Weather Events to Climate Change', *Irish Independent*, 26 July 2021.

18 Jon Williams, 'How RTÉ News Is Covering Climate Change', RTÉ News, 27 July 2021.

19 Donogh Diamond, 'Existence of Climate Change Never at Issue', *Irish Examiner*, 7 April 2014.

20 'How Much Will Fighting Climate Change Cost Ireland?', *Prime Time*, RTÉ, 3 December 2015.

21 'Climate Change the Cause of Last Year's Horrific Flooding in Midleton', RTÉ News, February 2024.

22 John FitzGerald et al., 'Medium-Term Review 2008–2015', Economic & Social Research Institute, May 2008.

23 Ciarán Hancock, 'Fitzgerald Admits ESRI "Totally Wrong" on Banking Collapse', *Irish Times*, 11 February 2015.

24 Richard Tol, 'Why Worry About Climate Change?', Economic & Social Research Institute, Quarterly Commentary, 21 April 2009.

25 Bob Ward, 'IPCC Corrects Claim Suggesting Climate Change Would Be Good for the Economy', *Guardian*, 17 October 2014.

26 Matt Ridley, 'Why Climate Change Is Good for the World', *Spectator*, 19 October 2013.

27 Sandy Trust et al., 'The Emperor's New Climate Scenarios', Institute and Faculty of Actuaries, University of Exeter, July 2023.

28 Frank Ackerman et al., 'Fat Tails, Exponents, Extreme Uncertainty: Simulating Catastrophe in DICE', *Ecological Economics* 69(8), 2010: 1657–65.

29 John FitzGerald, 'Nobel Laureate Shows Failure on Carbon Tax Policy Likely to Be Costly', *Irish Times*, 12 October 2018.

30 Pierre Jacques et al., 'The European Green Deal Requires a Renewed Economic Modelling Toolbox', Letter, Earth4All, 16 February 2024.

31 'WMO: 2015 Likely to Be Warmest on Record, 2011–2015 Warmest Five Year Period', World Meteorological Organization, 25 November 2015.

32 Joint Committee on Climate Action, 'Climate Change: A Cross-Party Consensus for Action', Houses of the Oireachtas, March 2019, p. 28.

33 'Climate Classroom with Chief Meteorologist Jeff Berardelli', WFLA News Channel 8, 23 January 2025.

34 Daniel Murray, '"Climate film industry is dying" – Eco Eye Producer to Shut Down After 20 Years', *Business Post*, 24 March 2024.

35 'Climate Crisis Needs "Everyday" Media Coverage, Committee Told', RTÉ News, 8 October 2024.

36 'Digital News Report Ireland 2023', Coimisiún na Meán, 13 June 2023.

37 Kevin O'Sullivan, 'Rising Tides: Ireland's Future in a Warmer World – Kills Off Myth We Are Among Small States With Minor Emissions', *Irish Times*, 27 March 2024.

38 Conor Skehan, 'Dublin at the Crossroads', Institute of International and European Affairs, 2009.

39 Frank McDonald, 'FG Adviser a Sceptic on Climate Change', *Irish Times*, 12 March 2011.

40 'Address by Taoiseach to UN Secretary General's Climate Change Summit, New York, 23 September 2014', RTÉ, 23 September 2014.

41 'Multinationals Urged to Consider Ireland', RTÉ News, 17 October 2014.

42 Paul Melia, 'Climate Change Is Not Our Priority – Taoiseach', *Irish Independent*, 1 December 2015.

43 Caroline O'Doherty, 'Eight Minutes at End of Leaders' Debate Shows Climate Is Nowhere Near Top of Election 2024 Agenda', *Irish Independent*, 27 November 2024.

44 Fiachra Ó Cionnaith, 'Half of Voters Think Govt Hasn't Done Enough on Climate Change', RTÉ News, 30 November 2024.

45 John Gibbons, 'RTÉ's Climate Week Must Be the Spark for a Coherent Action Plan', *Irish Times*, 18 November 2019.

46 'Climate Actions Work', Government of Ireland, 3 January 2025.

47 J. Ray Bates, 'Deficiencies in the IPCC's Special Report on 1.5 Degrees', Global Warming Policy Foundation, 2018.

48 Gavin Schmidt, 'Bending Low with Bated Breath', realclimate, 22 December 2018.

49 'Irish Agriculture and Climate Change – the Good News!', Irish Climate Science Forum, August 2024.

50 'Must Try Harder – Education Publisher called to Book', An Taisce press release, 4 September 2016.

51 Email from Folens to author, 18 November 2015.

52 Emma O'Kelly, 'Folens Revises Chapter on Global Warming', RTÉ News, 31 August 2016.

Chapter 8: A New Food and Land Vision

1 Nithiya Streethran et al., 'ClimAg: Multifactorial Causes of Fodder Crises in Ireland and Risks Due to Climate Change', EPA Research Report no. 464, 2018, p. 6.

2 Ibid.

3 Liam Lysaght, 'IPBES & Ireland's Biodiversity Crisis', National Biodiversity Data Centre website.

4 'National Strategy for Horticulture, 2023–2027', Department of Agriculture, Food and the Marine, 2023.

5 Tamara Fitzpatrick, 'Just 60 Field Vegetable Growers Left in the Country – Bord Bia Director', *Irish Independent*, 17 October 2023.

6 'Keelings – Case Study', Nature Rising website.

7 Ellie Howard, 'Victory Gardens: A War-Time Hobby That's Back in Fashion', BBC News, 26 May 2020.

8 'Franklin D. Roosevelt – Statement Encouraging Victory Gardens', The American Presidency Project, 1 April 1944.

9 Ellie O'Byrne, '"It's Amazing for Mental Health": Cork Community Gardeners Dig in as State Loses Plot', *Irish Examiner*, 25 February 2023.

10 'Cereal Crops', Teagasc website.

11 Teagasc Technology Foresight 2030, 'Irish Crop Production: Current Situation, Future Prospects to 2030 and Development Needs', Teagasc website.

12 E. J. Sheehy, 'The Future of Agriculture in Ireland', *Studies: An Irish Quarterly Review* 27(107), 1938: 455–66.

13 'Statistics Explained – Developments in Organic Farming', Eurostat, June 2024.

14 John P. Reganold and Jonathan M. Watcher, 'Organic Agriculture in the Twenty-First Century', *Nature Plants* 2, 2016: 15221.

15 John Gibbons, 'Pesticides – the Global Hazard That Is Harming Farmers' Health Here and Now', *Irish Examiner*, 5 October 2022.

16 'Pesticide Industry Lobby's Reckless Assault on Biodiversity and Health', Corporate Europe Observatory, 19 November 2023.

17 J. M. Bertolote et al., 'Deaths from Pesticide Poisoning: Are We Lacking a Global Response?', *British Journal of Psychiatry* 189(3), 2006: 201–3.

18 Joe Mag Raollaigh, 'Appeal to Reduce Pesticide Use to Protect Water Supplies', RTÉ News, 26 April 2024.

19 'Entrepreneur Marie Martin on How She Invented the Safe Scrub Sprayer', *Late Late Show*, RTÉ, 1 February 2019.

20 Francisco Sánchez-Bayo and Kris A. G. Wyckhuys, 'World-wide Decline of the Entomofauna: A Review of Its Drivers', *Biological Conservation* 232, April 2019: 8–27.

21 Tara Cornelisse et al., 'Elevated Extinction Risk in Over One-Fifth of Native North American Pollinators', *Proceedings of the National Academy of Sciences* 122(14), 2025: e2418742122.

22 Laura Reiley, 'Cutting-Edge Tech Made This Tiny Country a Major Exporter of Food', *Washington Post*, 21 November 2022.

23 Tom Vorstenbosch et al., 'Famine Food of Vegetal Origin Consumed in the Netherlands During World War II', *Journal of Ethnobiology and Ethnomedicine* 13, 2017: 63.

24 www.koppert.co.uk/crop-protection/biological-pest-control/.

25 'WUR Is Best Agricultural University in the World for Seventh Consecutive Time', Wageningen University & Research, 6 April 2022.

26 Toby Sterling, 'Gas Crisis Hits Dutch Greenhouses', Reuters, 8 September 2022.

27 Tom Levitt, 'Netherlands Announces €25bn Plan to Radically Reduce Livestock Numbers', *Guardian*, 15 December 2021.

28 Paul Tullis, 'Nitrogen Wars: The Dutch Farmers' Revolt That Turned a Nation Upside-Down', *Guardian*, 16 November 2023.

29 Sophie Tanno, 'Trump and Le Pen Backed These Dutch Farmers – Now They've Sprung an Election Shock', CNN, 19 March 2023.

30 'Who Will Feed Us? The Industrial Food Chain vs. the Peasant Food Web', ETC Group, 2017.

31 Leah H. Samberg et al., 'Subnational Distribution of Average Farm Size and Smallholder Contributions to Global Food Production', *Environmental Research Letters* 11(12), 2016: 124010.

32 David Pimentel, 'Soil Erosion: A Food and Environmental Threat', *Environment Development and Sustainability* 8(1), 2006: 119–37.

33 Chris Arsenault, 'Only 60 Years of Farming Left if Soil Degradation Continues', *Scientific American*, 5 December 2014.

34 William J. Sutherland et al., 'A Horizon Scan of Global Biological Conservation Issues for 2024', *Trends in Ecology & Evolution* 39(1), 2024: 89–100.

35 Xavier Poux and Pierre-Marie Aubert, 'An Agroecological Europe in 2050: Multifunctional Agriculture for Healthy Eating', IDDRI Study no. 9, September 2018.

36 'Local Food Policy – Putting Food Back into the Community', Talamh Beo website.

37 'Members of the Teagasc Authority', Teagasc website, April 2025.

38 Monica Caparas et al., 'Increasing Risks of Crop Failure and Water Scarcity in Global Breadbaskets by 2030', *Environmental Research Letters* 16(10), 2021: 104013; Toshihiro Hasegawa et al., 'Evidence for and Projection of Multi-Breadbasket Failure Caused by Climate Change', *Current Opinion in Environmental Sustainability* 58, October 2022: 101217.

39 Emmet Byrnes and Declan Little, 'A History of Woodland Management in Ireland: An Overview', Native Woodlands Scheme Information Note no. 2, Woodlands of Ireland.

40 'Irish Forests – a Brief History', Forest Service, Department of Agriculture, Fisheries and Food, 2008.

41 'Our Story', Coillte website.

42 Caren Jarmain et al., 'Creating and Managing Forests for Carbon from an Irish Perspective', *Irish Forestry Journal* 78(1&2), 2024: 11–53.

43 Ronan McGreevy, 'Storm Éowyn's €500m Toll on Irish Forestry Revealed by Satellite Imagery', *Irish Times*, 15 March 2025.

44 'Fire Back in Favour as Upland Farm Management Tool', *Irish Examiner*, 30 January 2014.

45 Declan Malone and Joan Maguire, '"Sickening" Swathe of Destruction at Gorta Dubha Gorse Fire', *The Kerryman*, 30 March 2022.

46 John Casey and Catríona Foley, 'Can We Assume All Fires Are Bad? Not Necessarily So', Teagasc, 14 January 2022.

47 'Irish Wildlife Trust Calls for All Burning of Uplands to Be Banned Under Any Revision of Wildlife Act', Irish Wildlife Trust, 8 January 2021.

48 Martin Merrick, 'Sustainable Use of Sheep Dips to Protect Efficacy', *Irish Farmers Journal*, 20 September 2023.

49 Colm Ryan, '43% of Sheep Farms Earned Family Farm Income of Under €5,000 in 2023', Agriland:.ie, 23 July 2024.

50 Noel Bardon, 'Two Measures Enough for €13/Ewe Payment – INHFA', *Irish Farmers Journal*, 9 October 2024.

51 Aisling O'Brien, 'No Change in Farming With New National Park – Noonan', Agriland:.ie, 22 April 2024.

52 Hanna Bijl, 'The Return of the Apex Predator in Europe', HAMS Online, 24 May 2023.

53 Lisa O'Carroll, 'EU to Rethink Conservation Status of Wolves After Numbers Surge', *Guardian*, 4 September 2023.

54 Harry McGee, 'Greens Call for Wolves to Be Reintroduced to Ireland', *Irish Times*, 1 October 2019.

55 Cormac McQuinn, 'Wolves Would All Be Shot if They Were Reintroduced in Ireland, Says Green Minister', *Irish Times*, 4 January 2024.

56 'Report: Rewilding and Climate Breakdown', Rewilding Britain website.

57 'Frequently Asked Questions', Farming for Nature website.

58 'All Ireland Pollinator Plan' and 'No Mow May', National Biodiversity Data Centre website.

59 F. Tanneberger et al., 'The Peatland Map of Europe', *Mires and Peat* 19(22), 2017: 1–17.

60 Catherine Brahic, 'Peatland Destruction Is Releasing Vast Amounts of CO_2', *New Scientist*, 11 December 2007.

61 John Gibbons, 'Destroying Bogs to Produce Uneconomical Energy', *Irish Times*, 13 August 2009.

62 Manus Boyle, 'EU Asked to Investigate Irish Subsidy for Biomass Burning With Peat', GreenNews.ie, 21 August 2018.

63 'From Carbon Source to Carbon Sink', Bord na Móna website.

64 Noel Bardon, 'Over €65m in Turf Cutting Compensation Paid Out', *Irish Farmers Journal*, 24 January 2024.

65 Padraic Fogarty, 'Continuing Failure by Ireland to Protect Peatlands', Irish Wildlife Trust, 29 September 2022.

66 'Joint Committee on Environment and Climate Action Debate', Houses of the Oireachtas, 16 February 2023.

67 Kevin O'Sullivan, 'Ireland Referred to EU Court for Failure to Protect Bog Lands and Curb Turf Extraction', *Irish Times*, 13 March 2024.

68 Marina Romanello et al., 'The 2024 Report of the *Lancet* Countdown on Health and Climate Change: Facing Record-Breaking Threats from Delayed Action', *The Lancet* 404(10465), 2024: 1847–96.

69 Niamh Griffin, 'Irish Children Face Shorter Life Expectancy and Malnutrition Caused by Climate Change', *Irish Examiner*, 30 October 2024.

70 'Farm Structure Survey 2023', Central Statistics Office, 17 December 2024.

71 Niall Sargent, 'Emissions-Heavy Feed Imports Soaring to Fuel Dairy Boom', TheJournal.ie, 5 October 2023.

72 'Global Food Policy Report 2016', International Food Policy Research Institute, 2016, p. 70.

73 Hannah Daly, 'Beef Is Not Sustainable, So Why Are We Subsidising It for Export?', *Irish Times*, 2 February 2023.

74 Rob Bailey et al., 'Livestock – Climate Change's Forgotten Sector', Chatham House Royal Institute of International Affairs, 2014, p. 22.

Chapter 9: Adapt or Die

1 'TRANSLATE: One Climate Resource for Ireland', Met Éireann website.

2 'Met Éireann publishes Ireland's new Climate Averages for 1991–2020', Met Éireann, 18 July 2023.

3 'Ireland Is the Wettest It Has Been in More Than 300 Years', ICARUS Climate Research Centre, Maynooth University, 28 March 2018.

4 John Gibbons, 'Climate Change Is Happening Right Here, Right Now', *Irish Times*, 1 September 2017.

5 'Marine Heat Wave 2023 – A Warning for the Future', Met Éireann website.

6 John Sweeney, 'Ireland Must Prepare Itself for the "Big Flood"', *Irish Times*, 4 November 2023.

7 'Major Weather Events' [1798–2024], Met Éireann website.

8 'Managing Flood Risk in Ireland', Office of Public Works, 12 October 2022.

9 Geoffrey Lean, 'UK Flooding: How a Yorkshire Town Worked With Nature to Stay Dry', *Independent*, 3 January 2016.

10 Colman O'Sullivan, 'Dubliners Help Their City Better Prepare for Floods', RTÉ News, 5 March 2024.

11 Olivia Kelly, 'Water System "in a desperate state", Says Uisce Éireann Chair', *Irish Times*, 28 March 2025.

12 Alexander Hall, 'The North Sea Flood of 1953', *Arcadia* 5, 2013.

13 'Room for the River', Rijkswaterstaat, Dutch Ministry of Infrastructure and Water Management website.

14 Tristan Baurick, 'The Dutch Are Giving Rising Rivers More Room. Should We Follow Suit?', Climate Central, 9 March 2020.

15 John Sweeney, 'Climate Change in Ireland: Science, Impacts and Adaptation', in D. Robbins et al. (eds), *Ireland and the Climate Crisis* (Macmillan, 2020), pp. 15–36.

16 Olivia Kelly, 'Council to Spend €500,000 on Lowering Clontarf Sea Wall', *Irish Times*, 9 January 2018.

17 Gary Ibbotson, 'Councillors Say Clontarf Flood Defences Are Urgently Needed', *Dublin People*, 27 May 2022.

18 'East Coast Railway Infrastructure Protection Projects – ECRIPP', Irish Rail website.

19 'Project Ireland 2040', Government of Ireland, 12 April 2025.

20 Daniel Murray, '"Risk" Coastal Protection Works on Dart Won't Work, Says Irish Rail Boss', *Business Post*, 14 May 2023.

21 Kelly De Bruin and Clement Kweku Kyei, 'Policy Brief on Economic Costs of Climate Change Impacts and Adaptation in Ireland', Climate Change Advisory Council Working Paper no. 32, September 2024.

22 Fernanda Terra Stori and Cathal O'Mahony, 'Coastal Climate Adaptation in Ireland', MaREI, University College Cork, 2021, p. 6.

23 Martim Mas e Braga et al., 'A Thicker Antarctic Ice Stream During the Mid-Pliocene Warm Period', *Communications Earth & Environment* 4, 2023: 321.

24 'The Flood Protection Gap', Central Bank of Ireland, October 2024.

25 Daniel Murray, 'Insurers May Withdraw Products Due to Climate Change – Ireland's Finance Commissioner', *Business Post*, 20 May 2024.

26 Alex Morrison, 'Climate Scenario Models in Financial Services Significantly Underestimate Climate Risk', University of Exeter, 4 July 2023.

27 Thomas Frank, 'Climate Change Is Destabilizing Insurance Industry', *Scientific American*, 23 March 2023.

28 René M. van Westen et al., 'Physics-Based Early Warning Signal Shows that AMOC Is on Tipping Course', *Science Advances* 10(6), 2024: eadk1189.

29 Kevin O'Sullivan, 'Collapse of Atlantic Ocean Current Could Turn Ireland's Climate into Iceland's', *Irish Times*, 16 February 2024.

30 John Gibbons, 'We Are Fast Approaching Climate Change Tipping Points', *Irish Examiner*, 24 April 2023.

31 Nathan Bindoff et al., 'Open Letter by Climate Scientists to the Nordic Council of Ministers', October 2024.

32 'A Stagnant Jet Stream Is Fuelling Intense Heat Worldwide. Could Climate Change Be to Blame?', *Yale Environment 360*, 19 July 2023.

33 James Hansen et al., 'Ice Melt, Sea Level Rise and Superstorms: Evidence from Paleoclimate Data, Climate Modeling, and Modern Observations that 2 °C Global Warming Could Be Dangerous', *Atmospheric Chemistry and Physics* 16(6), 2016: 3761–812.

34 'Over One Billion at Threat of Being Displaced by 2050 Due to Environmental Change, Conflict & Civil Unrest', Institute for Economics & Peace, 9 September 2020.

35 John Gibbons, 'Zambians Brace for Water Shortage Despite Recent Rainfall', *Guardian*, 12 March 2020.

36 'Zambia CO_2 Emissions 2011–2022', Trading Economics website.

37 Paul O'Callaghan, *Defence Forces Review 2024* ('Is Climate Change a Driver of the Far-Right?', in Óglaigh na hÉireann, Irish Defence Forces, January 2025), pp. 19–20.

38 William J. Ripple et al., 'The 2024 State of the Climate Report: Perilous Times on Planet Earth', *BioScience* 74(12), 2024: 812–24.

39 Richard Fisher, 'BBC Future at Hay Festival: How to Think Longer-Term', BBC News, 5 July 2019.

40 John Gibbons, 'Like Sweden and Wales, We Need a Ministry of the Future', *Business Post*, 10 June 2023.

Index